Special Topics in Particle Technology 2018

Topical at the 8th World Congress on Particle Technology

Orlando, Florida, USA
22 - 26 April 2018

ISBN: 978-1-5108-6970-7

Printed from e-media with permission by:

Curran Associates, Inc.
57 Morehouse Lane
Red Hook, NY 12571

CURRAN ASSOCIATES INC. proceedings.com

Some format issues inherent in the e-media version may also appear in this print version.

Printed by Curran Associates, Inc. (2018)

AIChE
120 Wall Street, FL 23
New York, NY 10005-4020

Phone: (800) 242-4363
Fax: (203) 775-5177

www.aiche.org

Additional copies of this publication are available from:

Curran Associates, Inc.
57 Morehouse Lane
Red Hook, NY 12571 USA
Phone: 845-758-0400
Fax: 845-758-2633
Email: curran@proceedings.com
Web: www.proceedings.com

TABLE OF CONTENTS

Some abstracts in this book refer to images that were not made available by the publisher

508549 Advanced Residue Fluid Catalytic Cracking (RFCC) Regenerator Analysis

Abstract:

A Residue Fluid Catalytic Cracking (RFCC) regenerator typically consists of two consecutive bubbling beds combusting coke on deactivated catalyst under partial burn (below stoichiometric) conditions. The RFCC regenerator may well be the largest piece of equipment in a refinery, so improvement in coke combustion efficiency in such a vessel can make it smaller and shorter and thus reduce capital investment for a new RFCC unit.

CFD analysis was employed to understand the flow patterns inside the regenerator and develop a more efficient design. A two-fluid model for gas and solids, coupled with the kinetic theory, and employing the EMMS drag force correlation were used to simulate the flow of air and particles and their respective residence time distribution in the regenerator. The new regenerator design features a compartmentalized second stage with individual air distributors supplying air to each compartment. Because of different air flow rates through these air distributors, the EMMS drag model was tuned differently for the flow in the inner and outer compartment to match the commercial observations from similar systems.

CFD results shows the radial pressure decreases at the bed interface from the regen center to the chamber wall. Residence time distributions of air showed a typical bubbling bed behavior corresponding to four CSTRs in series per compartment, while those of catalyst followed a two-CSTR behavior. Compartmentalization led to a more efficient use of oxygen in the regenerator which now required smaller fluidized bed volumes to sufficiently regenerate catalyst. The latter directly translates into the capital costs of the vessel.

508076 A Numerical Study on Steric Crowding Effect in Inertial Microfluidics

Abstract:

In square microchannels, flowing particles can laterally migrate and focus into streamlines due to hydrodynamic inertia. This inertial focusing has attracted intense attention for promising capability of passive particle manipulation without applying any external field. One aspect that remains unexplored is the effect of particle concentration on focusing behaviour. Although seemingly efficient for the manipulation, high particle concentration (1-3 v/v%) has been suggested to create steric crowding effect that might impede complete particle focusing. However, the concentration effect has not been studied because this steric crowding effect involves high-throughput flow on the micro scale of highly condensed suspension flow. This nature considerably defies experimental exploration. In this study, we numerically study the concentration effect on the inertial focusing behaviour for highly condensed suspension flowing in square microchannels. The numerical method is the discrete element method coupled with the direct numerical simulation for particle-fluid interaction [1]. This method is referred to as the DEM-DNS method. We find that complete focusing behaviour disappears above ~3 v/v%. It is also suggested that this steric crowding effect arises when flowing particles are within close distance in high concentration, and consequently, particle rotation is significantly reduced as a result of viscous effect of inter-particle hydrodynamics. This mechanism could also imply the essential role of particle rotation in formation of distinct streamlines found in square microchannels whose detailed mechanism has not been fully explained. Finally, the DEM-DNS method is shown to be a promising tool for inertial microfluidic analysis.

[1] H. Udono and M. Sakai, â€œA numerical study on dynamic inertial focusing of microparticles in a confined flowâ€, *Granular Matter* (accepted)

507165 Assessment of Powder Cohesion Strength Using Shear Cell and Warren Spring Cohesion Testers

Abstract:

The characterization of powders and bulk solid materials is extremely important to ensure uniform product quality. The flow obstructions in both large and small scales lead to process down time, bin collapse, compromised aerolisation of dry-powder inhalers (DPIs), etc. Many shear cell instruments have been designed to characterize the flow behaviour of particulate materials. However, most of these devices are limited to work at linear boundary of the yield locus, i.e., at higher normal stress loads. The cohesion index values are obtained by extrapolation of the yield locus. This limitation compromises the characterization of cohesiveness at low normal stresses (< 100 Pa). This study compares the cohesion index of common food and pharmaceutical powder materials obtained from a shear cell and Warren Springs cohesion (WSC) instrument. The powders analysed were starches (from wheat, potato and corn), microcrystalline cellulose (MCC) and wheat gluten. Volume mean diameter ($D_{4,3}$) of the particles ranged from 118.47 ± 5.26 μm for gluten to 27.71 ± 1.41 μm for corn starch. For MCC, the cohesion index using WSC at 0.1 kPa was 6.55 ± 0.02, approximately, four times the value for wheat starch (1.59 ± 0.28). Intermediate values were found for corn starch samples (0.84 ± 0.02) and gluten samples (0.53 ± 0.18), whereas potato starch samples presented the lowest cohesion index (0.38 ± 5×10^{-3}). The WSC values will be compared with the shear cell values to characterize the difference in cohesion index due to different measurement techniques. At completion of this study, the cohesion index values of common powder materials, at low stress levels, will be established.

505928 Simulating Radiation and Conduction in an Aggregate Dryer Using Coupled CFD-DEM

Abstract:

Granular heat transfer is an important mechanism in many industrial applications. While convective heat transfer in multiphase systems is common in some industrial applications radiation and conduction are dominant modes and must be considered. An application in which all modes are critical is an aggregate dryer in a hot mix asphalt plant. For efficient mixing, the aggregate material must be completely dried in order to be coated by the liquid asphalt cement binder. This process is accomplished in a counter flow drum heated by a direct fire burner with internal metal slats (flighting) to facilitate heat transfer. In the region immediately surrounding the flame, special flights are designed to shield the drum from radiation. Direct observation of the dryer is impossible and thus simulation gives the ability not only to monitor the behaviour of the material, but also to optimize the flight design for increased drying efficiency and mitigate to thermal damage to the drum from the heat of the burner flame. In this study coupled CFD-DEM was used to investigate different flight designs in order to optimize both heat transfer to the aggregate particles and thermal shielding of the drum shell. Radiation from the flame to the particles and drum and flights was included as well as conduction between particles and geometry. The results of the simulation have been corroborated by field observations.

506150 Modeling of Simultaneous Particle Attrition and Pipe Wear

Abstract:

Conveying particles by a carrier fluid in pipelines has considerable advantages, such as continuous transport, energy efficiency, safety, and dust control. However, two major drawbacks exist with this process: particle attrition, and pipeline erosion that could defect the product quality and cause system failure. For this reason, it is essential to minimize these problems by optimizing the design of the system prior to construction and working with optimized operating conditions. But in practice, the optimization process by prediction of either erosion or breakage is complicated and time consuming tasks.

There are different approaches for coping with this problem: purely empirical, combined computational tools with experimental measurements or fully theoretical modeling. Due to the increase of computational power in the recent years, the most common way to predict these phenomena became computational fluid dynamics coupling with another tool to assess the breakage or erosion. An example for such approach, which is widely used, is the coupling of Discrete Element Method (DEM)-Computational Fluid Dynamics (CFD) simulations with a discrete erosion model. The major advantage of this approach is it's the straight forward way for saving the collision data (particle impact velocity, angle and specific location) to serve as the input for the erosion model. However, this combination is usually limited to compute the erosion rate at a single pipeline component, which is typically the most vulnerable part in the conveying system, such as contraction or bends, instead of a full system analysis. The reason behind this limitation is the high computational requirements for a full system with huge number of particles. When simulating breakage, the same limitation repeats because of the large amounts of fragments which eventually preventing a full scale prediction of the particle size distribution (PSD).

Regardless of its size, a full scale optimization for the dynamic parameters, e.g. particle velocity and pressure drop, can be done with simplified models. Those models compute the two-phase dynamics with zero dimension (no dependency on the pipe length) or one-dimensional (computing the dynamic parameters in terms of the pipe axial direction) models. The advantage of this method is of course its computation time, nonetheless, 3D effects such as turbulent eddies and swirling flow cannot be accurately computed via this method and therefore 3D evaluation of the particles' trajectories is not possible. The consequence of this limitation is the disability to compute explicitly particle-particle and particle-wall interactions, and therefore it cannot express the input needed to breakage or erosion models.

Recently, we developed two models that are combining the advantages of the CFD-DEM and the one-dimensional two phase flow simulations. By doing so, they are capable of predicting erosion or breakage in long-range conveying systems in short computation time. The particle attrition model is solving conventional 1D balance of mass, momentum and energy to compute the two-phase flow dynamics. These set of equations are coupled with the one dimensional breakage algorithm (ODBA) for predicting the alternation of the particles' size distribution along the pipeline. The ODBA is combining three essential elements: (i) a tracking procedure, (ii) particle breakage characterization, and (iii) collision characteristics. Tracking is done only on a representative mass which is defined by a chosen number of particles. The particle breakage characteristics defines if a particle would break or not upon impact, and what would be the sizes of its fragments if it does breakage. Those characteristics were adopted from empirical based measurements. The last piece of the puzzle is the collision characteristics that expresses the frequency, location, velocity, and angle of the impact. These characteristics were derived from CFD-DEM data into statistical distribution functions. This complete model allows to forecast the PDS at every desired place in the conveying system, while considering fatigue and p-p as well as p-w interactions.

The second model, for erosion prediction, which is known as one-dimensional erosion model (ODEM), is based on a similar principal like the ODBA. The only fundamental difference from the ODBA is replacing the 'particle breakage characterization' with a discrete erosion model that is able to compute the wear depth due to the particle impact. This model needs the input from the collision characteristics in order to give the desired output: erosion rate map of the entire conveying system. Yet, both the ODBA and ODEM have a concealed assumption that is very widely used, but rarely investigated. The energy loss from an impact is attributed only to the investigated phenomena, i.e., when a simulation is after prediction of erosion it is regularly assumed that the particle is rigid, and therefore the energy loss due to the impact is because the surface deformed plastically, or micro cracks were forming. However, many times this is not the case, because the particles are damaged and attrition is well distinguished.

In the current study we address this issue by presenting a new concept for dealing with combined erosion and breakage. The first step toward this cause was to quantify the collision characteristic in terms of the operating conditions, and to prove that those characteristic can be equivalent in both erosion and breakage simulations. The main difference between breakage and erosion CFD-DEM simulations, which the data was collected from, is the materials and the operating conditions. When examining the comparison between breakage or erosion collision characteristics, it turned out that the collision characteristics could be derived as non-dimensional parameters, and they share similar trends for different conveying conditions. One distinct feature that was found is that the non-dimensional impact velocity could be derived as the actual impact velocity divided by the median impact velocity. In addition, the median impact velocity was found to correlate well with the axial velocity of the particles, which means that if the axial velocity is known then so does the median impact velocity. The essence of such analysis is that the actual impact velocity could be predicted in terms of the conveying conditions, without depending on the results from CFD-DEM simulations.

Another contribution of this work is presenting for the first time the methodology of combination between breakage and erosion. When observing at an impact of a discrete particle with a surface, both parties are damaged. The purpose of our analysis is to provide a scheme that recognize which of the bodies was damaged, and what mechanism of erosion or breakage occurs. Furthermore, this recognition allows to quantify with a proper model how much the surface was eroded, if any, or was the particle fractured and broken or not.

Implementing this methodology to the combined ODBA and ODEM, with the analyzed collision characteristics allows a complete modeling of simultaneous breakage and erosion. This features a novel modeling to a phenomenon that was not addressed before for the best of our knowledge.

508101 Spatially-Averaged Models for Large-Scale Gas-Solid Flows

Abstract:

Fluidized beds are widely used in a variety of industrially important processes. During the last decades the analysis of the hydrodynamics or the efficiency of fluidized beds through numerical simulations has become increasingly common. During the last decades the analysis of the hydrodynamics or the efficiency of fluidized beds through numerical simulations has become increasingly common, where the two-fluid model (TFM) approach has proven to provide fairly good predictions of the hydrodynamics of gas-solid flows.

However, due to computational limitations a fully resolved simulation of industrial scale reactors is still unfeasible. In our previous study (Schneiderbauer 2017), we have presented a spatially-averaged two-fluid model (SA-TFM), which enables the coarse grid simulation of dense large-scale gas-solid flows. However, these averaged TFM equations require constitutive models for the residual correlations appearing due to averaging. On the one hand, the unresolved part of the gas-solid drag force has been derived by employing a series expansion to the microscopic drag coefficient and on the other hand the Reynolds-stress like contributions are closed similar to Boussinesq-approximation in single phase flows. The subsequent averaging of this linearized drag force reveals that the unresolved part of the interphase momentum exchange is a function of the turbulent kinetic energies (TKE) of both, the gas and solid phase, and the variance of the solids volume fraction. Closure models for these quantities have been derived from first principles.

In contrast to TFM, parcel based approaches, such as MP-PIC (O'Rourke & Snider 2010) and DDPM (Cloete & Amini 2016), have become quite popular recently to access the numerical simulation of large scale gas-solid flows. However, similar to TFM computational limitations restrict such simulations of large scale gas-solid flows to coarse grids (for the continuous phase) as well (Li et al. 2012; Benyahia & Sundaresan 2012). In this paper, we show that the constitutive relation derived for the SA-TFM apply also to the parcel based approaches since the parcel approach represents a Lagrangian discretization of the solids transport and momentum equations. We refer this coarse grained parcel approach as SA-DDPM.

Finally, the SA-TFM and SA-DDPM models are validated for three different fluidization regimes: First, we apply these coarse grid models to the NETL challenging problem (Panday et al. 2014), where the flow of Geldart A and B particles in a riser section of a circulating fluidized bed has been studied. Second, the predictions of SA-TFM in the turbulent fluidization regime is examined (Zhu et al. 2008). Finally, a bubbling fluidized bed is studied (Zhu et al. 2008). The numerical results obtained on coarse grids demonstrates that the SA-TFM and SA-DDPM reveal nearly identical results and further fairly good agreement of pressure drop, solids volume fraction as well as velocity profiles. Thus, the results proof the SA-TFM and SA-DDPM are applicable to a wide range of particle diameters and different fluidization regimes.

References:

Benyahia, S. & Sundaresan, S., 2012. Do we need sub-grid scale corrections for both continuum and discrete gas-particle flow models? *Powder Technology*, 220, pp.2–6. Available at: http://linkinghub.elsevier.com/retrieve/pii/S003259101100595X [Accessed April 17, 2012].

Cloete, S. & Amini, S., 2016. The dense discrete phase model for simulation of bubbling fluidized beds: Validation and verification. In A. Soldati & C. Marchioli, eds. *Proceedings of the 9th International Conference on Multiphase Flow*. Florence, Italy.

Li, F. et al., 2012. MP-PIC simulation of CFB riser with EMMS-based drag model. *Chemical Engineering Science*, 82, pp.104–113. Available at: http://dx.doi.org/10.1016/j.ces.2012.07.020.

O'Rourke, P.J. & Snider, D.M., 2010. An improved collision damping time for MP-PIC calculations of dense particle flows with applications to polydisperse sedimenting beds and colliding particle jets. *Chemical Engineering Science*, 65(22), pp.6014–6028. Available at: http://linkinghub.elsevier.com/retrieve/pii/S0009250910005075 [Accessed March 17, 2012].

Panday, R. et al., 2014. Challenge problem: 1. Model validation of circulating fluidized beds. *Powder Technology*, 258, pp.370–391. Available at: http://dx.doi.org/10.1016/j.powtec.2014.02.010.

Schneiderbauer, S., 2017. A spatially-averaged two-fluid model for dense large-scale gas-solid flows. *AIChE Journal*.

Zhu, H. et al., 2008. Detailed measurements of flow structure inside a dense gas-solids fluidized bed. *Powder Technology*, 180(3), pp.339–349.

509396 Dynamics and Energetics for a Vertically Stirred Mill: Validation of a Discrete Element Method (DEM) Model Via Positron Emission Particle Tracking (PEPT)

Abstract:

8th World Congress on Particle Technology

Dynamics and Energetics for a Vertically Stirred Mill: validation of a Discrete Element Method (DEM) Model via Positron Emission Particle Tracking (PEPT).

Domenico Daraio[**], Hugh E. Stitt [*] Alessio Alexiadis [**], Andrew Ingram[**], and Michele Marigo [*]

* Johnson Matthey Technology Centre - Chilton Site - Billingham (UK)

** School of Chemical Engineering - University of Birmingham – Birmingham – (UK)

e-mail: Domenico.daraio @matthey.com,

ABSTRACT

Vertically stirred attritor mills are widely used for comminution in numerous industrial applications ranging from the ceramics, paint, pharmaceutical and mineral industries. This type of mill can be utilised for both dry or wet milling in order to achieve ultrafine grinding.

The comminution energy is transferred by a motor driven impeller to the grinding media to achieve the wanted particle size reduction. Unfortunately, not all the energy is effectively used in the reduction process of the product, a large part of it is dissipated by energy loss during the collisions and friction among the grinding media, in the forms of heat and noise. The full understanding for both the media motion and the media ball-media ball, media ball-impeller and media ball-grinding chamber interactions are the key features that need to be analysed in order to improve the efficiency of the equipment and/or the product quality. In addition, often the design and the scale-up are developed on an empirical basis due to the lack of understanding of how the energy is really transferred from the grinding media to the powder and what are the dominant mechanisms for the media motion.

The current work addresses these limitations with a more mechanistic approach, adopting the Discrete Element Method (DEM LIGGGHTS, by DCS computing, Linz -Austria), to reveal the exchange of stresses and force distribution during the milling operation under a variety of conditions (impeller speed, media filling, shaft design, impeller configuration). Positron Emission Particle Tracking measurements, a non-invasive technique, has been utilised for model validation by investigating the grinding beads dynamics inside a 1.4lt lab attritor mill under similar conditions to the DEM model. A Matlab routing has been adapted and modified to fit the specific case, to produce azimuthally averaged velocity and occupancy distributions of the PEPT data.

A sensitivity study has been carried out for four DEM input parameters describing the particle properties (Young's Modulus, static friction coefficient, rolling friction coefficient, coefficient of restitution) using a matrix of simulations according to the principle of the Design of Simulations (DoS). 16 simulations were necessary to understand the role of each parameter and their impact on the simulation results was assessed against two responses: the power consumption coming from the experiments (global validation of the DEM model) and the local velocity calculated using the PEPT raw data (local validation of the DEM model).

It has been found that the frequency distribution of velocity obtained from the PEPT experiments, (tracking a single particle for an extended time), is not equivalent to the DEM-velocity distribution based on the entire particle population. However, the DEM model is capable of capturing the main features of the grinding media motion in term of torque on the impeller and velocity of the particles. In fact, the DEM model predicts well the local grinding media velocity at different radial positions along the attritor mill height, allowing more trustful considerations on the collision energetics. Inside the attritor mill have been identified and classified four distinct area where the particle motion is controlled by different mechanisms: rolling and/or static friction between layers of particles, gravity, impacts between the particles. More important the extension of these area will affect the efficiency of the equipment and /or product quality. Combining the information of the residence time and the particle velocity, the death area inside the equipment have been mapped and they have been minimised by optimising the impeller configuration.

For a more quantitative evaluation of the different milling conditions, the DEM data have been time and space averaged using a novel coarse-graining post-processing tool (Particles Analytics Ltd, Edinburgh, UK). Continuous fields of velocity, kinetic pressure, solid fraction and number of contacts were computed and used as starting point for a re-design of the impeller.

REFERENCE

[1] R. W Rydin, D. Maurice, T. H Courtney, *Milling dynamics: Part I. Attritor dynamics: Results of a cinematographic study,* Metallurgical Transactions A, Volume 24A, January 1993, Pages 175-185

[2] Priya R. Santhanam, Alexandre Ermoline, Edward L. Dreizin, *Discrete element model for an attritor mill with impeller responding to interactions with milling balls*, Chemical Engineering Science, Volume 101, 20 September 2013, Pages 366-373, ISSN 0009-2509

[3] N.S. Weerasekara, M.S. Powell, P.W. Cleary, L.M. Tavares, M. Evertsson, R.D. Morrison, J. Quist, R.M. Carvalho, *The contribution of DEM to the science of comminution*, Powder Technology, Volume 248, November 2013, Pages 3-24, ISSN 0032-5910,

[4] Jankovic, A., 2003. Variables affecting the fine grinding of minerals using stirred mills. Minerals Engineering, 16, 337-345

[5] Fabio Chiti, Serafim Bakalis, Waldemar Bujalski, Mostafa Barigou, Archie Eaglesham, Alvin W. Nienow, Using positron emission particle tracking (PEPT) to study the turbulent flow in a baffled vessel agitated by a Rushton turbine: Improving data treatment and validation, Chemical Engineering Research and Design, Volume 89, Issue 10, 2011, Pages 1947-1960, ISSN 0263-8762,

[6] I. Govender, M.S. Powell, An empirical power model derived from 3D particle tracking experiments, Minerals Engineering, Volume 19, Issue 10, August 2006, Pages 1005-1012

509585 Structural and Mechanical Characterization of the Calendaring Process of Lithium-Ion Electrodes Via Discrete Element Method Simulations

Abstract:

Lithium-ion batteries are widely used as electrochemical power sources for mobile telephones, personal computers, cameras and other modern-life appliances. They are also remarkably successful in the electric power vehicle market due to their long live cycles and high-rate capabilities.

However, as lithium-ion batteries appear to be gaining in popularity, there is still room to improve their performance and durability. In fact, the relevance of electrode structural stability and mechanical integrity has been already pointed out by several research groups [1]. Lithium-ion battery electrodes consist of porous composite materials (active material, conductive additives and binder) coated on a substrate. Microstructural characteristics of such particulate coatings, namely porosity and particle size distribution are of major importance in determining bulk properties. For instance, the performance of a lithium-ion battery can be significantly enhanced by selecting the adequate electrode porosity or by adjusting the correct amount of binder. Concerning their mechanical aspects, it is well-known that stress generation within the electrodes is one of the main causes for capacity fade and eventual failure of lithium-ion batteries. For this reason, mechanical instabilities, including particle fragmentation, structure disintegration and fracturing, loss of contact between the electrode and the current collector or plastic deformation are still a major subject of extensive research activities [2].

Within this framework, the present study proposes a discrete element method (DEM) procedure to create particulate coatings with defined product characteristics and to give insight into the influence of microstructure with regard to the mechanics. Operating the open source software LIGGGHTS with additional programming code, particulate structures can be computer-generated taking into consideration a specific particle size distribution, bulk porosity and coating thickness. This process is automatized and fulfilled by an iterative adjustment of input parameters. The application of this effective designed tool is put into practice for the numerical generation of lithium-ion electrode structures. A key component of such coatings is the binder, which ensures the cohesion throughout the particulate network and the adhesion between the coating and the substrate. The role of the binder in the mechanics of the system is therefore essential and must also be considered for the simulations. Bearing this fact in mind, a Hertzian-bond contact model was also developed in this study in order to capture the elasto-plastic behavior of the electrode. This contact model combines both particle and binder stiffness by computing bonds between particles under certain conditions with regard to interparticle distance and particle radii.

In this study, different lithium-ion battery electrodes were experimentally produced as well as numerically generated in order to validate the implemented procedure. Furthermore, nanoindentation experiments and corresponding simulations were carried out with the aim of calibrating and validating the DEM contact model. Nanoindentation is a useful technique that coupled with DEM simulations, contributes to get a deeper look into the mechanical properties of the electrodes. This outcome provided an essential comprehensive understanding of how the interaction among the electrode components may affect the mechanical and electrochemical properties of the whole structure.

The validated computer-generated structures were subsequently used to simulate the calendaring process. This relevant manufacture step consists of compacting the electrodes with two rolls directly after coating and drying. Calendaring has been proved to have a substantial impact on the pore structure and, as a result, on the electrochemical performance of the lithium-ion battery. Hence, it is of extreme importance to comprehensively control this process in order to achieve targeted battery performances [3]. Within this work, an additional electrode was produced and calendered at different line loads by varying the gap size between rolls. Respective simulations enabled to provide detailed knowledge concerning the influence of this process on stress generation by calculating the macroscopic stress associated to each calendered structure. Moreover, simulations brought the advantage of contributing to a better understanding regarding the evolution of the microstructure and, in particular, the gradual change of particle contacts. This feature is directly related to the conductivity of the electrode, which is an essential aspect with regard to the performance of the battery. Ultimately, this research may be valuable not only to get deeper knowledge into the manufacture process but also to optimize electrode design with the aim of achieving specific battery requirements

[1] S. W. Peterson, D. R. Wheeler, Direct Measurements of Effective Electronic Transport in Porous Li-Ion Electrodes, Journal of The Electrochemical Society, 161 (14) A2175-A2181 (2014)

[2] A. Mukhopadhyay, B. W. Sheldon, Deformation and stress in electrode materials for Lithium-ion batteries, Progress in Materials Science 63 58-116 (2014)

[3] C. Meyer, H. Bockholt, A. Kwade, Characterization of the Calendering Process for Compaction of Electrodes for Lithium-Ion Batteries, Journal of Materials Processing Technology 249 (2017)

507254 Evaluating the Influence of Dry Granulation Processing Variables on the Rheological Properties of Granules

Abstract:

Granulation is often employed in the production of pharmaceutical formulations for oral solid dose delivery. It is most often required when constituent materials in a formulation have poor flow properties, enabling these components to be combined to improve process performance and ensure content uniformity.

Dry granulation, based around the use of a roller compactor and integral mill/screen, can have significant advantages over more common wet granulation techniques. Dry granulation can reduce processing costs and can also be used with moisture/solvent/heat sensitive materials. Unlike a wet granulate, the dry granulate does not require additional drying and milling before encapsulation or tabletting.

Dry granulation is not suitable for all types of material, however, there is little research into which properties make a formulation suitable for this process. In addition, there is little to indicate which process parameters produce optimal granulate quality in order to achieve interruption free processing and high quality final products.

In this study an FT4 Powder Rheometer® (Freeman Technology, Gloucestershire, UK) was used to demonstrate how processing parameters influence the properties of a placebo formulation granulate. The granulate was produced using a roller compactor (MiniPactor, Gerteis Maschinen+Processengineering AG, Switzerland) on which the roll gap and compaction force were varied. The dynamic flow, bulk and shear properties of the resulting granulates were then evaluated using the FT4 Powder Rheometer.

The data clearly demonstrates that several rheological properties, known to influence die filling efficiency and Critical Quality Attributes (CQAs) of the resulting tablets, could be controlled by changes in roller compactor settings.

This study emphasises the importance of assessing powder properties using techniques that closely simulate process conditions and how it is possible to identify Critical Process Parameters to optimise a dry granulation process and ensure high final product quality.

507489 Assessment of the Homogeneity of an Alumina Powders Mixture Using Rheological Parameters

Abstract:

This study has been carried out in order to improve the comprehension of powder mixing in the nuclear fuel manufacturing process. Thus, a binary mixture of two Alumina (Al_2O_3) powders, GE15 and CR6, was used to simulate the oxides powders actually involved in the process. Different mixing technics and parameters were tried and compared in terms of homogeneity of the mixtures produced. The concept of intensity of segregation [1] was used for assessing mixture homogeneity through a sampling protocol and classical statistical analysis tools [2]. One of the difficulties of working with these compounds is that, due to the similarities between CR6 and GE15 powders (same chemical composition, same colour, etc.), the direct measure of the composition of a given sample is not possible. Since homogeneity of a mixture is assessed from the variation of the composition between samples, it is important to find an indirect way to measure the composition of an Alumina powders mixture. Therefore, different correlations between measurable parameters (specific surface area, Sauter diameter, cohesion and flow function coefficient) and the composition of the mixtures were found empirically allowing indirect measurement of the composition of a given sample.

The four studied parameters: specific surface area (, measured by nitrogen adsorption using BET method), Sauter diameter (, measured by laser granulometry, dry transportation with air at 0.1 bar), cohesion and flow function coefficient (and , measured by FT4 shear test at 9 kPa on a 10 mL cell), were compared in terms of precision, relevance and feasibility.

It appears that the homogeneity assessment methods involving rheological parameters (cohesion and flow function coefficient) show many advantages compared to the other methods. Not only are they the most accurate correlations found but both parameters are measured simultaneously, meaning that it is possible to apply both methods on the same samples while all the tests (N_2 adsorption, laser granulometry and FT4) are destructive. Those two methods also allow the estimation of the homogeneity at a more interesting scale (around 7 g of powder), indeed, specific surface area and Sauter diameter methods give a homogeneity at a scale of less than one gram which is not relevant for the end-use properties of the final product.

[1] P. Danckwerts, The definition, measurement and some characteristics of mixtures, Applied scientific research. Section A, 3 (1952) 279-296

[2] S. Massol-Chaudeur, H. Berthiaux, S. Muerza, et J. Dodds, A numerical model to identify the structure of a high-dilution powder mixture, *Powder Technol.*, 128 (2002) 131-138

508516 Characterising the Grain Shape: In Search of Size Independent Shape Descriptors

Abstract:

Granular materials contain a large number of grains that differ in size and shape. While the characterisation of the size distribution is typically done using sieve analysis, the characterisation of the shape of the grains remains a difficult task and is one of the major bottlenecks in the characterisation of the granular material. In this study, we attempt to characterise the shape of three different types of grains commonly used in iron and steel making industry, namely pellet, sinter and iron ore lump. While the pellet particles can be considered nearly spherical in shape, imposing the spherical shape on the iron ore lumps and highly irregular sinter particles seems a very poor choice. We consider two different size ranges of pellet particles and four different size ranges for sinter and iron ore lumps. Imaging of the grains is done using an iPhone 7 Plus camera and sufficient number of grains of each type in all the size ranges are considered. We perform the image analysis and obtain various boundary parameters (such as projected area, minor and major axis, perimeter, etc.) of the grains using MATLAB image processing toolbox. In addition, we define two additional boundary parameters to characterise the deviation of the projected area from that of a circle and calculate their values for all the different type of grains considered in this study. Using these length scales, we define a total of nine different shape descriptors and obtain the distribution of these shape descriptors for each type of grain for all the size ranges. Interestingly, for a given type of grains, the frequency distribution curves for these nine shape descriptors turn out to be nearly independent of the size range of the grains. This suggests that the chosen shape descriptors indeed represent properties related to the shape of the grains alone, irrespective of the size of the grains considered.

We investigate the correlation among these shape descriptors by evaluating the values of the correlation coefficients between each pair of the shape descriptors using the Pearson method, Kendall method and Spearman method. Our results indicate that out of the nine shape descriptors considered in this study, five shape descriptors are strongly correlated with each other and only the remaining four shape descriptors need to be considered in characterising the shape of the grains. Out of these four uncorrelated shape descriptors, two shape descriptors, namely the aspect ratio of the grains and the root of form factor, have been considered important by researchers in previous studies. The two other shape descriptors proposed in this study, which are used to characterise the deviation of the projected grain shape from that of a circle, are not measured previously to the best of our knowledge. We compare these four shape descriptors for the pellets, sinter and iron ore lumps having the same sieve size range and show that the shape descriptors for these three different types of grains are different.

We next calculate the values of these four shape descriptors for relatively larger coke particles using 2d image analysis. We also perform 3d image analysis of some of these coke particles and calculate the values of the analogous shape descriptors in 3 dimensions. We investigate if the identified 2d shape descriptors (obtained from 2d imaging) can be correlated to the analogous 3d shape descriptors (obtained from 3d imaging) and explore if it is possible to use the 2d imaging technique to estimate the 3d shape descriptors, thus eliminating the need of 3d imaging.

508543 Investigating Mechanical Properties of Surfactant Films at the Solid-Liquid Interface Using AFM

Abstract:

Corrosion is a well-known problem in the oil and gas sector and also in many other industries. Surfactant based inhibitors have been used for corrosion control in oil pipelines but their inhibition mechanism is still not fully known. Our group is studying surfactant corrosion inhibition mechanism by various characterization techniques (AFM, Electrochemical and Spectroscopic techniques). A better understanding of corrosion inhibition mechanism will lead to the design and development of not only cost-effective but also greener surfactants system for corrosion inhibition purposes.

Atomic force microscopy (AFM) is an effective tool to study the barrier properties of surfactant films at the solid-liquid interface. Determination of mechanical and topographical properties of single and mixed surfactant coatings by atomic force microscopy (AFM) can delineate the role of physical vs. electrochemical barrier by such films in the corrosion inhibition process. . So far, the AFM data, obtained using a closed cell, indicate the effect of surfactant concentration and that of hydrocarbon chains interactions on the mechanical properties of cationic surfactant films. Ultimately, integration of the AFM and electrochemical results will reveal the correlation between the surfactant structures at the solid-liquid interface and their corrosion inhibition efficacy.

508545 Deposition Rate Consequences of the Formation of Multi-Spherule Cluster Aggregates in Gases "Role of Momentum Shielding^{TM}"

Abstract:

For engineering flow systems characterized by particulate suspensions in gases, will the formation of *cluster aggregates* (CAs) in the mainstream accelerate deposition rates on immersed targets or confining walls ---or is aggregate formation more likely to suppress capture rates? Recent research on the drag properties of large cluster aggregates (e.g., stimulated in part by the Sorensen (2011) review) has opened the door to answering these important mechanism- (and pressure-) dependent questions. Sub-100 nm "primary" spherules, which may find themselves in even tenuous 'fractal-like' N-spherule clusters, can take advantage of what we call "momentum shielding"---analogous to the drag advantages experienced by birds who elect to fly in formation. We show here that the extent of this "momentum shielding" can be conveniently quantified (in a dimensionless function we call: $S_{mom}(N;Kn_1;$aggregate structure)), and used to predict the deposition-rate consequences of aggregation in aerosol flow systems when the cluster deposition mechanism is dominated by either: i) isothermal convective diffusion (C-D), ii) thermo-phoresis (TP-) or iii) inertial impaction (II). As expected, we show that for any type of aggregate structure (including 'fractal-like'), momentum shielding increases with cluster size (e.g. N) and gas pressure (via the reciprocal Knudsen number based on spherule radius R_1). We also show how S_{mom} can be calculated for the limiting case of a N>>1 random close-packed-D_f=3 cluster ---which is nearly gas impermeable. For fractal-like aggregates, including DLCAs (D_f=1.8) and RLCAs (D_f=2.1) we have recently shown that the abovementioned function: $S_{mom}(N, Kn_1; D_f)$ is actually sufficient to make useful engineering estimates of CA Brownian diffusion coefficients (Rosner and Tandon (2017b), thermophoretic diffusivities (Rosner and Tandon (2017a) and aggregate "stopping times" (Rosner and Tandon (2017c)----each in terms of their previously known values (D_1, $\alpha_T D_1$, and $t_{p,1}$) for isolated individual spherules in the same environment.

This information now enables instructive engineering calculations of the deposition-rate consequences of spherule aggregation to form large cluster aggregates-----i.e. at the same mainstream spherule volume fraction we can now determine how much aggregation will increase or decrease the total spherule acquisition rate on an immersed target. We present here representative results for each of the abovementioned particle transport mechanisms----i.e.: i) isothermal convective-diffusion (C-D), ii) thermo-phoresis (TP-) or iii) inertial impaction (II)----- including a brief account of the consequences of dealing with Kn1-dependent coagulation-'aged' aggregate size distributions (ASDs) in the mainsteam. We summarize our current re-examination of the canonical problem of inertial impaction on a circular cylinder in Re>>1 crossflow (Rosner and Tandon (2017c), dealing now with CA-"projectiles" but in the absence of spherule loss from the much larger target. We conclude by briefly outlining our planned work on the efficient "inversion" of sampled cluster-ASD-data to reconstruct, using the abovementioned $S_{mom}(N,Kn_1;D_f$ information, what ASD must have existed in the mainstream---at least in the absence of appreciable supercritical aggregate break-up.

References:

1. Rosner, D. E. and P. Tandon (2017a). Knudsen Transition Effects on the Thermophoretic Properties of Fractal-like Aggregates; Implications for Thermophoretic Sampling of High-Pressure Flames, Aerosol Science and Technology (in press).

2. Rosner, D. E. and P. Tandon (2017b). Aggregation- and Knudsen Transition Rarefaction- Effects on Particle Mass Deposition Rates by Convective-Diffusion, Thermophoresis or Inertial Impaction; Consequences of Multi-spherule 'Momentum Shielding', Aerosol Science and Technology (in press).

3. Rosner, D. E. and P. Tandon (2017b). Mainstream Aggregation and Polydispersity Effects for Inertially-Dominated Deposition Rates on Cylinders (or Leading 'Edges') in Crossflow"---- Consequences of 'Momentum Shielding' for the Impaction of Large, "Compact" (2<D_f<3) Multi-spherule Aggregates, Aerosol Science and Technology (prepared for submission, August 2017).

4. Sorensen, C. M. (2011). The Mobility of Fractal Aggregates: A Review, Aerosol Science and Technology. 45: 765-779.

508609 Experimental and Numerical Studies on the Thermal Diffusivity of Packed Powder Beds

Abstract:

Several industrial processes, such as calcination and drying, involve thermal treatment of particles. In order to achieve efficient processes and good quality products, the thermal properties of the particles and that of the bulk powder must be well defined. Previously, some models, such as the Krischer and Maxwell-Eucken models, have been proposed to relate particle thermal properties to that of the bulk powder. Though these models have been shown to work well, it is not very clear how to account for changes in the packing density, moisture content, and arrangement of particles. In addition, the particle thermal properties of most materials are not readily available, which limits the application of these models. To get a better understanding of the relationship between the particle thermal properties and that of the bulk powder and also to quantify the effects of packing density and moisture content on the bulk powder thermal properties, we run several experiments and discrete element method (DEM) based simulations. In both the experiments and simulations, a pipe is filled with particles where the temperature of the pipe is kept constant. We measured the temperature change in the powder bed and compute the actual thermal diffusivity of the powder bed. We tested several types of powders and glass beads in the experiments. The packing fraction, the moisture content, and size of particles were varied in the experiments. In the simulations, in addition to the packing fraction and particle size, the thermal properties of the particles were also varied. In this talk, we will present the details of the experimental and simulation results.

509556 Determination of Size, Size Distribution and Refractive Index of Artificial and Biological Microparticles

Abstract:

For the approval of clinical studies, it is mandatory to validate the production process of artificial oxygen carriers by reliably specifying the properties of the microparticles used like the newly developed blood substitute HbMP-700 (1-3). In particular, besides shape, size and size distribution, the amount of hemoglobin (Hb) and methemoglobin (metHb) content is essential.

We present a method based on the simultaneous measurement of collimated and diffuse transmission spectra in the visible and ultraviolet range and numerical analysis to infer the wavelength-dependent refractive index, size and size distribution of microparticles.

For the validation of the method, we used aqueous suspensions of monodisperse polystyrene microspheres of known concentration, determined by flow cytometric measurements. The related data analysis is carried out for synthetic and experimental data. For homogeneous spheres, Mie theory is applied to compute the solution for the scattering of a plane electromagnetic wave by a particle and to calculate the extinction cross section C_{ext}, i.e. the shadow cast by the particle as a function of the vacuum wavelength λ, the particle size, and the complex refractive index $n(\lambda)$. For dilute suspensions of particles, the measured quantity is an ensemble-averaged extinction cross section C_{ext} . This average is obtained by integrating over the particles' size distribution, which may not be known a priori. From the resulting spectral extinction cross sections, the complex refractive index and the mean and standard deviation of the size distribution can be inferred by means of nonlinear least-squares optimization. In order to reduce the number of free parameters of the problem to a feasible amount, $n(\lambda)$ is represented in a set of harmonic-oscillator-like resonance functions. The refractive index of polystyrene microspheres obtained is in good agreement with literature values for the bulk material, the mean and standard deviation of the distribution of particle *size* D are determined with an uncertainty in the range of a few nanometers. For example, for microspheres of mean diameter mean(D) = 2μm and distribution width std(D) = 9.4nm the uncertainty of the mean is $u\{\text{mean}(D)\}$ = 1.2nm and the uncertainty of the distribution width amounts to $u\{\text{std}(D)\}$ = 0.4nm.

In a second step, we applied the method to characterize HbMP-700 microparticles, approximately featuring a prolate spheroidal shape with 700nm length and 400nm width. Besides shape, size and size distribution, the amount of hemoglobin (Hb) and methemoglobin (metHb) content is essential. The relative concentrations of oxyHb, deoxyHb, metHb are accessible from spectral extinction and diffuse transmission measurements, since the different Hb variants exhibit different wavelength-dependent refractive indices. The method to infer size and refractive index described above can be applied to these non-spherical particles, since the deviation of their light scattering properties from those of spheres turns out to be small. The results based on the optical method will be compared to alternative procedures for HbMP particle characterization with respect to their size and size distribution, i.e. electron microscopy, centrifugation and dynamic light scattering. In addition, x-ray spectrometry will be used to detect the ratio between Fe^{2+} (oxy/dexyHb) and Fe^{3+} (metHb) as well as the iron concentration.

References

1) Novel hemoglobin particles--promising new-generation hemoglobin-based oxygen carriers. Bäumler H, Xiong Y, Liu ZZ, Patzak A, Georgieva R. Artif Organs. 2014 Aug;38(8):708-14. doi: 10.1111/aor.12331.

2) Nonvasoconstrictive hemoglobin particles as oxygen carriers. Xiong Y, Liu ZZ, Georgieva R, Smuda K, Steffen A, Sendeski M, Voigt A, Patzak A, Bäumler H. ACS Nano. 2013 Sep 24;7(9):7454-61. doi: 10.1021/nn402073n.

3) Hemoglobin-based oxygen carrier microparticles: synthesis, properties, and in vitro and in vivo investigations. Xiong Y, Steffen A, Andreas K, Müller S, Sternberg N, Georgieva R, Bäumler H. Biomacromolecules. 2012 Oct 8;13(10):3292-300. doi: 10.1021/bm301085x.

517131 FlowCam Nano ® Provides Counts, Sizes and Images of Nano-and Microparticles: Application to a Therapeutic Protein Pumping Study

Abstract:

Sub-visible particle characterization is becoming a more important method in assessing drug substance and drug product stability. However, the current USP protocol does not require any characterization of particles below 10 Î¼m. This greatly hampers our ability to properly assess protein stability as a function of particulate formation and obscures most sub-visible particles and various morphological properties.

In this study, we assess the capabilities of the new FlowCam Nano from Fluid Imaging Technologies, in conjunction with a pumping study that focuses on particulate formation generated during fill operations. Initial experiments show that the new FlowCam Nano can generate highly resolved images of nanoparticulates, allowing the simultaneous assessment of particulate morphology for nano- and microparticulates. This is extremely useful, as currently unpublished data (for the pumping study) show that as a function of post-pumping agitation, a significant increase in microparticle concentration coincides with a significant depletion of the nanoparticle concentration. This suggests that agglomeration (and probably not nucleation) is the more likely mechanism for larger microparticulate formation.

The ability to characterize the nano- and microparticulate morphology simultaneously, will allow us to better assess the cause of particulate formation (such as observing the transition of nanoparticles to microparticles and seeing if specific morphological properties of larger nanoparticulates are conserved in microparticulates). In addition, the FlowCam Nano is also sensitive enough to assess differences in the colloidal stability of a drug product in different formulations. During filling operations, and accelerated degradation studies, FlowCam Nano showed the differences in size distributions of IVIG in PBS and 0.25 M glycine buffer. Since IVIG is less colloidally stable in PBS, the size distribution varies significantly more per size bin and the overall mean effective spherical diameter is also ~ 600 percent greater after 24 hours of agitation in comparison to the size distribution data of IVIG in glycine. Thus, Flow Cam Nano has capabilities that are very useful in assessing drugs in the substance (e.g. optimizing drug formulations) and product (e.g. assessing particulates post fill & finish operations) phases.

Numerical and experimental estimation on the normal and tangential capillary bridge force adhered to two spheres

Kazuo Murase*, Keisuke Arai, Takato Ootsuka, Daiki Sakamoto and Futa Egawa,
Department of Applied Chemistry, Chuo University
1-13-27, Kasuga, Bunkyo-ku, Tokyo,1128551, Japan
**hmurase@kc.chuo-u.ac.jp*

Introduction

Liquid bridges extensively occur as a binder liquid adhered to solid surfaces in a variety of engineering applications such as crystal growth, ink-jet printing, material processing and powder technology. Especially the strength of adhesion and cohesion of liquid bridges adhered to particles contribute to the consolidation of an agglomerate in wet granular materials. Estimation of the attractive or repulsive force between particles is important in controlling kinetics of agglomeration and handling of granular materials. Characteristics of a liquid bridge e.g. the equilibrium bridge geometry [1], stability [2], static and dynamic binding force [3-6], rupture energy [7,8], and wetting hysteresis [9] etc., have been experimentally and numerically investigated.

The liquid distributions within a particles bed are classified into four types of categories, i.e. pendular, funicular, capillary and slurry states, according to the liquid contents [10]. The static normal capillary force at pendular liquid state have elaborately been investigated as Fisher [11], toroidal model [8] and the exact solution of Yung-Laplace equation [12]. The dynamic normal capillary forces [13] are experimentally estimated with lubrication theory. The tangential force of a liquid bridge [14] also contributes to the primary factor for the granular consolidation with the shear stress of wet granular materials but have not yet substantially been investigated.

This work aimed to estimate experimentally and numerically the normal and tangential capillary force of a static and dynamic liquid bridge adhered to two spheres. The asymmetry static liquid bridge with the shear stress demand for the exact solution of Young-Laplace equation for the equilibrium liquid drop including the effect of gravity.

1. Theory

1.1 Formulation of a asymmetry static liquid bridge

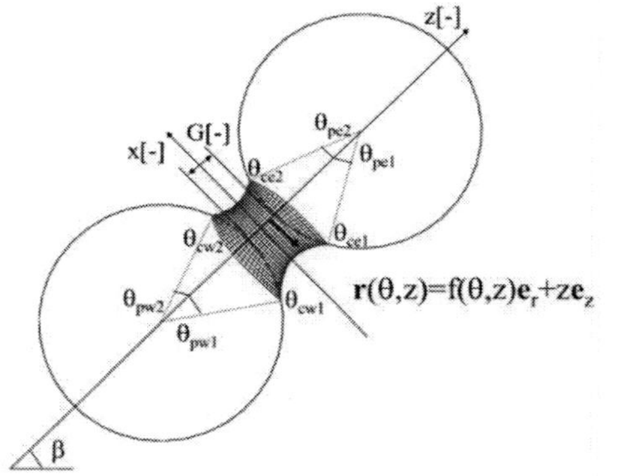

Figure 1 Geometry of a liquid bridge adhered to two spheres with arbitrary inclination in cylindrical coordinate system.

The asymmetry static bridge shape demands for the exact solution of the Young-Laplace equation due to the limit of the toroidal approximation. A prevalent geometry of a liquid bridge adhered to two spheres is drawn in Figure 1. The dimensionless position vector of a liquid surface in cylindrical coordinates is described as $\mathbf{r}(\theta,z) = f(\theta,z)\mathbf{e_r} + z\mathbf{e_z}$ divided by spherical radius, R', which $\mathbf{e_r}$, $\mathbf{e_z}$ are unit basis vectors in r-,z-directions. The symbol (') means the dimensional value. Fig.1 indicates the two dimensional front view with four half-filling (θ_{Pw1}, θ_{Pw2}, θ_{Pe1}, θ_{Pe2}) and contact angles (θ_{Cw1}, θ_{Cw2}, θ_{Ce1}, θ_{Ce2}).

The unit outward normal vector of a bridge surface, $\mathbf{n}_L$, is presented as follows [15],

$$\mathbf{n}_L \equiv \frac{r_\theta \times r_z}{|r_\theta \times r_z|} = \frac{f\mathbf{e_r} - f_\theta \mathbf{e}_\theta - ff_z \mathbf{e}_z}{[f^2(f^2+f_z^2)+f_\theta^2]^{1/2}} \quad \ldots(1)$$

,where low suffixes (r,θ,z) describe partial derivatives.

The dimensionless effective potential energy, $E=E'/\sigma R'^2$ with Lagrangian multiplier, $K=K'/\sigma R'^2$, on the constrained condition of a constant volume, $V=3V'/4\pi R'^3$, is described as follows [16],

$$E = \int_{-\pi/2}^{\pi/2}\int_{z_{\min}}^{z_{\max}} 2\left[f^2\left(f^2+f_z^2\right)+f_\theta^2\right]^{1/2} d\theta dz + \frac{Bo}{2}\int_{-\pi/2}^{\pi/2}\int_{z_{\min}}^{z_{\max}} (z\sin\beta + x\cos\beta) f^2 d\theta dz + K\left[V - \frac{1}{2}\int_{-\pi/2}^{\pi/2}\int_{z_{cp1}}^{z_{cp4}} f^2 d\theta dz\right] \quad \ldots(2)$$

Bond number (Bo) is a dimensionless number defined by $(\rho_L-\rho_G)gR'^2/\sigma$, where is the ratio of gravitational force to surface tension force equally as Etovoes number (Eoe).

The integral domain of a bridge surface is determined by the trapezoidal section of the upper and lower surface lines, [$z_{w2}<z<z_{e2}$ at $\theta=\pi/2$, $z_{w1}<z<z_{e1}$ at $-\pi/2$] and contact lines, [$z_{w1}<z<z_{w2}$, $z_{e1}<z<z_{e2}$ at $-\pi/2<\theta<\pi/2$] with four contact points settled by four half-filling angles ($\theta_{Pe1},\theta_{Pe2},\theta_{Pw1},\theta_{Pw2}$) in Fig. 1. Two contact lines on the upper and lower spheres were approximated by ellipse lines with parametric aspect ratios, which pass through respectively pairs of two contact points settled with half filling angles, ($\theta_{Pw1},\theta_{Pw2}$) and ($\theta_{Pe1},\theta_{Pe2}$). Filling angles, ($\theta_{PW}(\theta)$, $\theta_{PE}(\theta)$) on the lower and upper spheres have variable properties along the circumferential direction. The formulation of a static bridge shape at equilibrium state results in the extremum problem, which corresponds to the condition that the first order variation equals to zero, $\delta E=0$.

1.2 Estimation of the capillary force of a liquid bridge

The dimensionless mean curvature (H) and contact angles (θ_c) of a bridge surface are respectively calculated by eqs. (3),(4) once the outline of a liquid bridge is numerically evaluated.

$$-2H = \nabla_s \cdot \mathbf{n}_L = \left[(\alpha^2 + f_\theta^2) - f(f^2 + f_\theta^2)f_{zz} - f(1 + f_z^2)f_{\theta\theta} + 2ff_z f_\theta f_{\theta z}\right]/\alpha^3 \quad \text{...(3)}$$
$$,where\ is\ \alpha^2 = f^2(1+f_z^2) + f_\theta^2$$

$$\theta_c(\theta) = \cos^{-1}(\mathbf{n}_s \cdot \mathbf{n}_L) \quad \text{...(4)}$$

,where contact angles also depend on the circumferential angle.

According to the Young-Laplace equation, the dimensionless pressure difference, $\Delta P= \Delta P'R'/\sigma$, across a liquid surface is related with the mean curvature of the surface.

$$\Delta P = (-2H) \quad \text{...(5)}$$

The capillary force on a liquid bridge adhered to two spheres must be evaluated only on the contact line as known by the boundary method due to unspecified neck radius of a bridge and asymmetric condition. Dimensionless averaged capillary forces, $F_{Ci} = F'_{Ci}/(2\pi R'\sigma)$ on the upper and lower sphere (i=1,2 respectively correspond to lower, upper sphere) are presented as follows by three terms, which correspond to the surface tension, hydrostatic pressure and buoyancy terms along the longitudinal angle (θ). The sign of the capillary force set to be positive for attractive direction.

$$F_{ci} = \frac{1}{\pi}\int_{-\pi/2}^{\pi/2} \sin(\theta_{Pi})\sin(\theta_{Pi}+\theta_{Ci})d\theta - \frac{1}{2\pi}\int_{-\pi/2}^{\pi/2} \sin^2(\theta_{Pi})\Delta P d\theta \mp \frac{2}{3}\mathrm{Bo}\cdot V_{bi} \quad \text{...(6)}$$

, which V_b is sphere volume submerged in a bridge

Finite Element Method is applied for the solution of the $\delta E=0$ equation. Isoparametric elements of a biquadratic polynomial shape function were selected for the integral domain and unknown radius. The set of non-linear algebraic equations for unknown radius at element nodes and Lagrangian multiplier was resolved by the Newton-Raphson method [9].

2. Experiment

Tensile strength of a static and a dynamic pendular liquid bridge

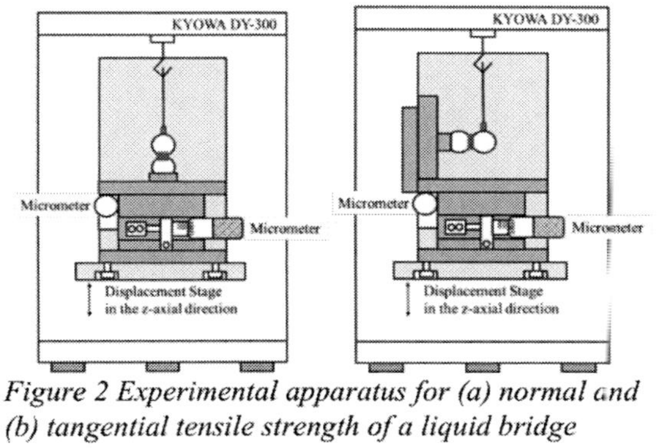

Figure 2 Experimental apparatus for (a) normal and (b) tangential tensile strength of a liquid bridge adhered to two spheres

Figure 2 shows the surface tensiometer (Kyowa interface Science Co., Ltd. DY-300) for measuring (a) the normal and (b) the tangential tensile strength of a water liquid bridge adhered to two stainless steel spheres (radius 3.0mm). The rod attached with one sphere is hanged up to the tensiometer and the other sphere fixed on the displacement stage is moved upward and downward by the tensiometer. The discrete and continuous displacement of the stage are performed for the normal and tangential deformation of a liquid bridge for the static and dynamic states. Two micrometers adjust the appropriate position of the lower spheres for the upper sphere. Water droplets are poured into the spheres aperture and substantially wetted by the displacement of the stage. The tensile strength at static and dynamic state and snapshots of liquid droplets by the digital camera and movie are recorded. The outlines of droplets are evaluated by a digitizer and estimate the half filling angles (θ_{Pe1},θ_{Pe2},θ_{Pw1},θ_{Pw2}) and contact angles (θ_{Cw1},θ_{Cw2},θ_{Ce1},θ_{Ce2}) at four contact points. The sphere configuration (radius, gap), the droplet volume, the half filling angles and the contact lines on spheres initiate parameters of numerical simulation.

3. Results and discussions

3.1 The normal tensile strength of a liquid bridge adhered to two vertical spheres.

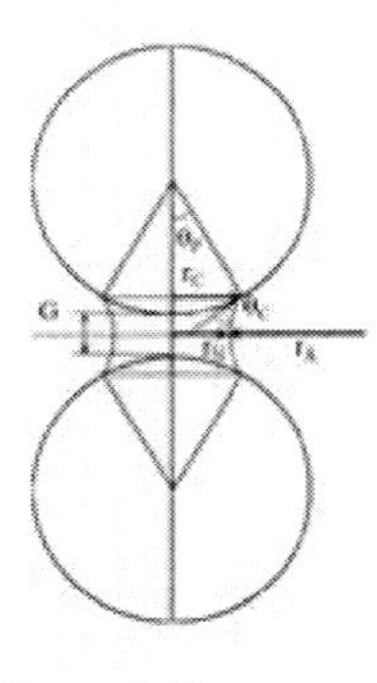

Figure 3 Geometry of toroidal model

The symmetric geometry of a liquid bridge adhered to two vertical spheres without gravity has been extensively investigated by toroidal approximation in Figure 3. The boundary method overestimates the capillary for of a liquid bridge rather than the gorge method but is feasibly and distinctly defined for the asymmetric bridge outline. Symmetric condition simplifies eq.(3) for the mean curvature and eq.(6) for the capillary force into eq.(7),(8) [7,16,17].

$$-2H = \nabla_s \cdot \mathbf{n}_L = \frac{1}{\alpha}\left(\frac{1}{f} - \frac{f_{zz}}{\alpha^2}\right), \textit{where} \quad \alpha^2 = 1 + f_z^2 \qquad \ldots(7)$$

$$F_{Boundary} = r_C \sin(\theta_{Pi} + \theta_{G}) - \frac{r_C^{\ 2}}{2}(\Delta P) \mp \frac{2Bo}{3} V_{bi}, \quad r_C = \sin\theta_P \qquad \ldots(8a)$$

$$F_{Borge} = r_N - \frac{r_N^{\ 2}}{2} \sin\theta_{cpi}(\Delta P) \mp \frac{2Bo}{3} V_{bi} \qquad \ldots(8b)$$

,where $r_N - r_A = \dfrac{A}{\tan(\pi/2 - \theta_P - \theta_C)} + \sin\theta_P, r_A = \left[A^2 + \{\sin\theta_P - (r_N - r_A)^2\right]^{1/2} A = G/2 + 1 - \cos\theta_P$

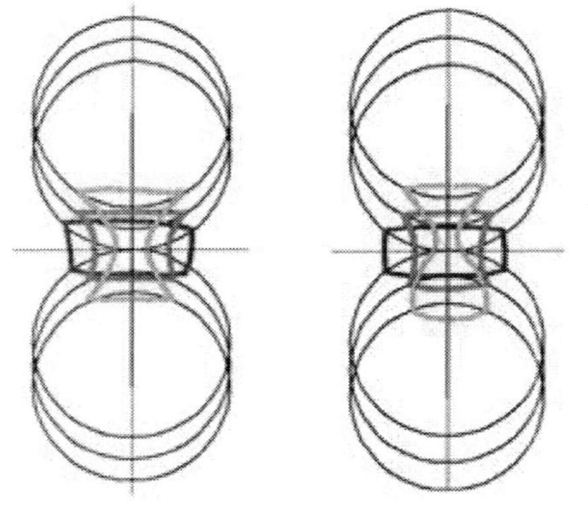

(a) Static state $G=0, 0.47, 0.91$ (b) Dynamic state $G=0, 0.59, 1.16$

Figure 4 Experimental droplet outlines at static and dynamic state, Bo=1.21, V=0.088

Figures 4 show symmetric outlines of liquid bridges around the vertical axis respectively at the static and dynamic state. Static state results in the shorter rupture gap and the larger dimensionless normal tensile strength from Figures 5a. Sphere gaps at the peak static capillary force depend on liquid volume. The contact angle on the upper sphere corresponds to the advancing contact angle at the higher Bo number due to gravity effect.

Fig. 4b indicates the downward transition of a liquid immediately before the rupture due to the wettability a liquid. The toroidal approximations over-estimate experimental data at the lower sphere gap but the sphere gaps at the peak dynamic capillary force depend on the velocity of stage displacement. The velocity reduces the peak value of dynamic capillary due to the downward transition of a liquid in Fig. 5b. The stepwise profiles of capillary force are due to the movement sensitivity of tensiometer for the stage displacement.

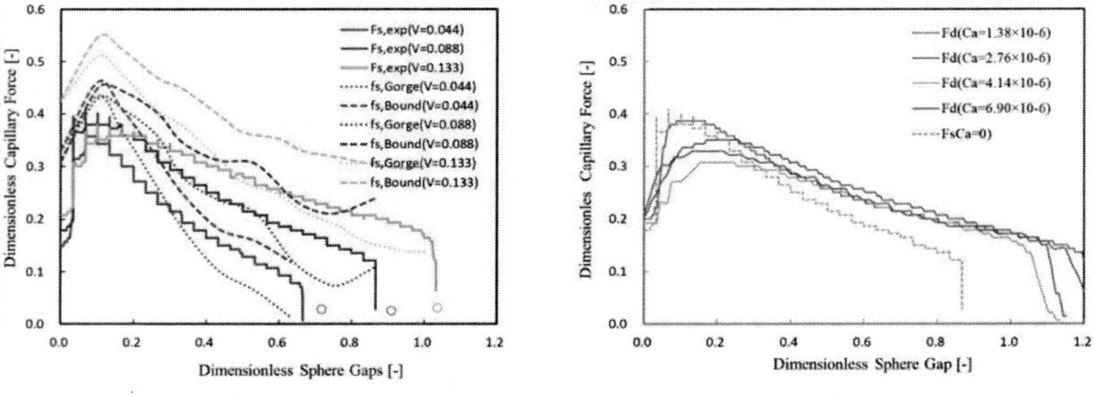

(a) at the static state (b) at the dynamic state

Figure 5 Effect of sphere gap on dimensionless capillary force

3.2 Estimation of boundary condition at contact line by LBM

The Young's equation describes the force balance among three interface tensions (solid-liquid, solid-gas and liquid gas) at three phase contact point and also defines the relationship between interface tension and contact angle. The solution of the Young-Laplace equation at the static state demands for one unknown boundary condition of either a contact angle or a contact line. The simulation for the wettability on three dimensional contact line demands for the elaborate process. Our experiment determines the contact lines on spheres under the constant liquid volume and is utilized for the boundary condition. Besides the perfect wettability the smooth transition of three phase contact lines is unexpected due to the advancing and receding contact angle. Lattice Boltzmann Method (LBM) is expected to estimate the boundary condition at static and dynamic state. The geometry of a liquid bridge is numerically estimated by FEM with half-filling angles predicted by 2 dimensional Shan & Chen model of LBM, which is used by D2Q9 model for transition of distribution function. The estimation for the liquid geometry by FEM has been verified. Figures 6 show the predicted liquid outlines by LBM and FEM. at a given liquid volume.

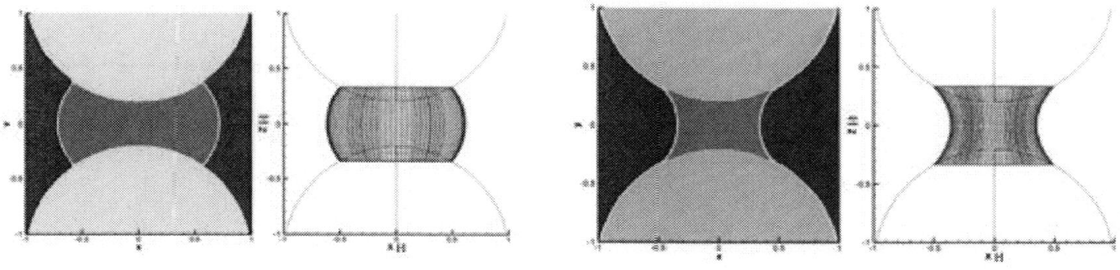

(a) Repulsive interparticle force (b) Attractive interparticle force

Figure 6 Snapshots of liquid geometries at static state by LBM and FEM

Figures 7 show snapshots of dynamic liquid outlines at a given liquid volume up to the rapture state by LBM. The transition of boundary conditions, i.e. filling and contact angles, is smoothly described. Gravity and imperfect wettability, i.e. advancing and receding contact angles, are excluded in Figs. 7 but the effect of gravity on the liquid outline is avairable.

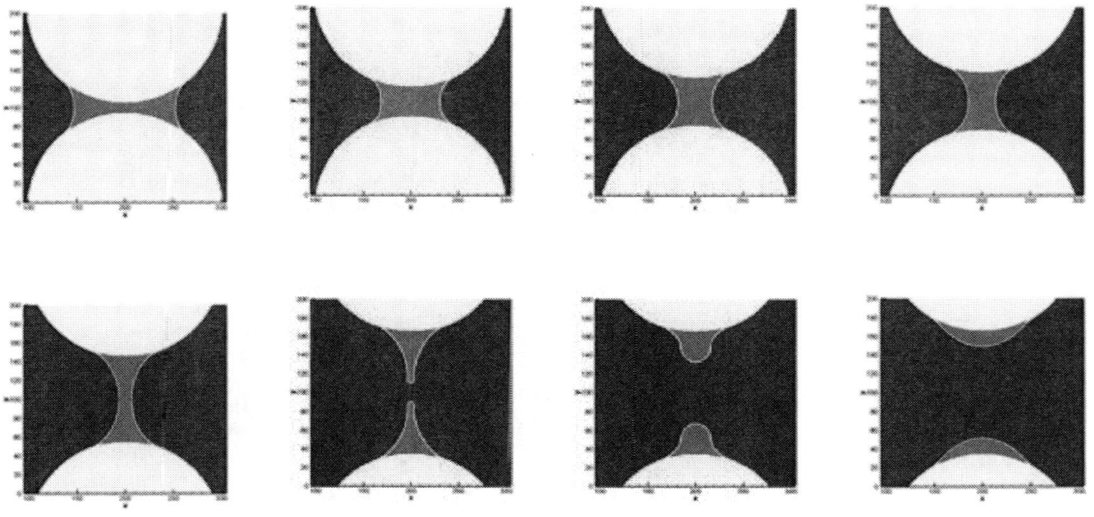

Figure 7 Snapshots of snapshots of dynamic liquid outlines at a given liquid volume

3.3 The tensile strength of a liquid bridge with shear deformation.

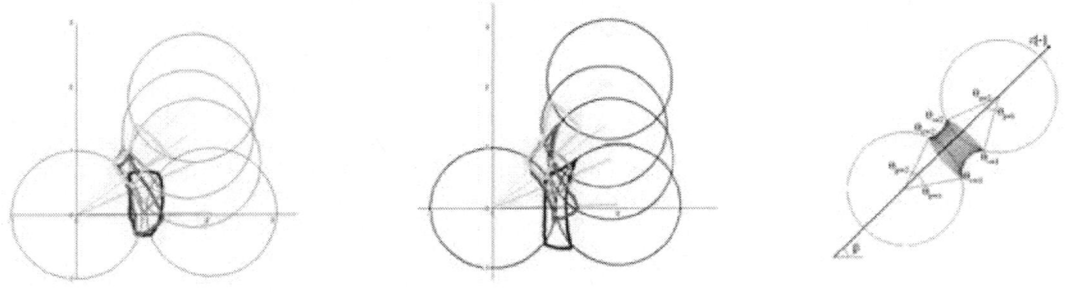

(a) Static state (G=0, 0.59, 0.98) (b) Dynamic state (G=0, 0.59, 0.98)

Figure 8 Liquid outlines at static and dynamic state (Bo=1.21, V=0.088)

Definitions of half-filling and contact angles.

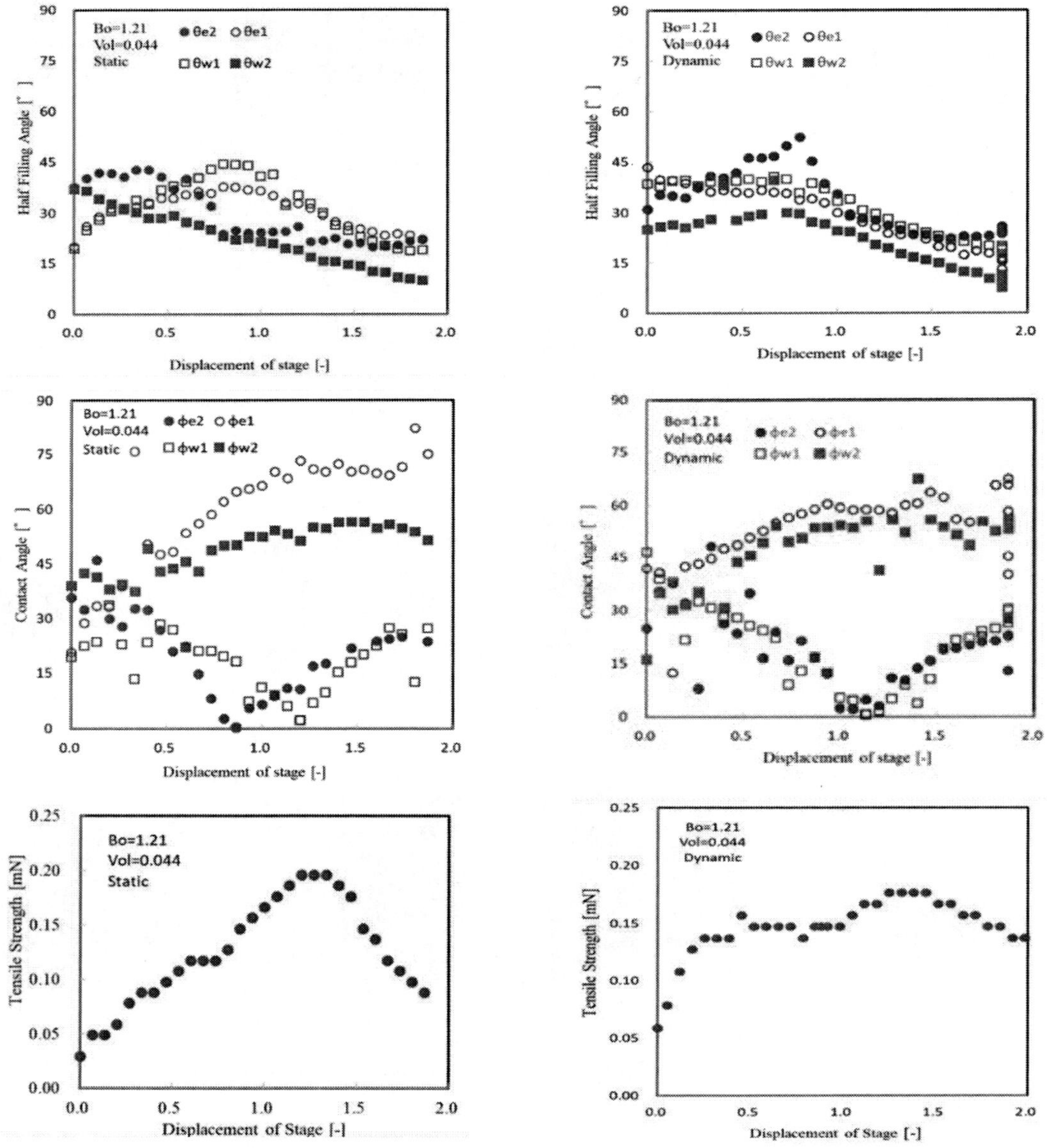

(a) Static state (b) Dynamic state

Figure 9 Profiles of half-filling angles, contact angles and tensile strength , Bo=1.21, V=0.088

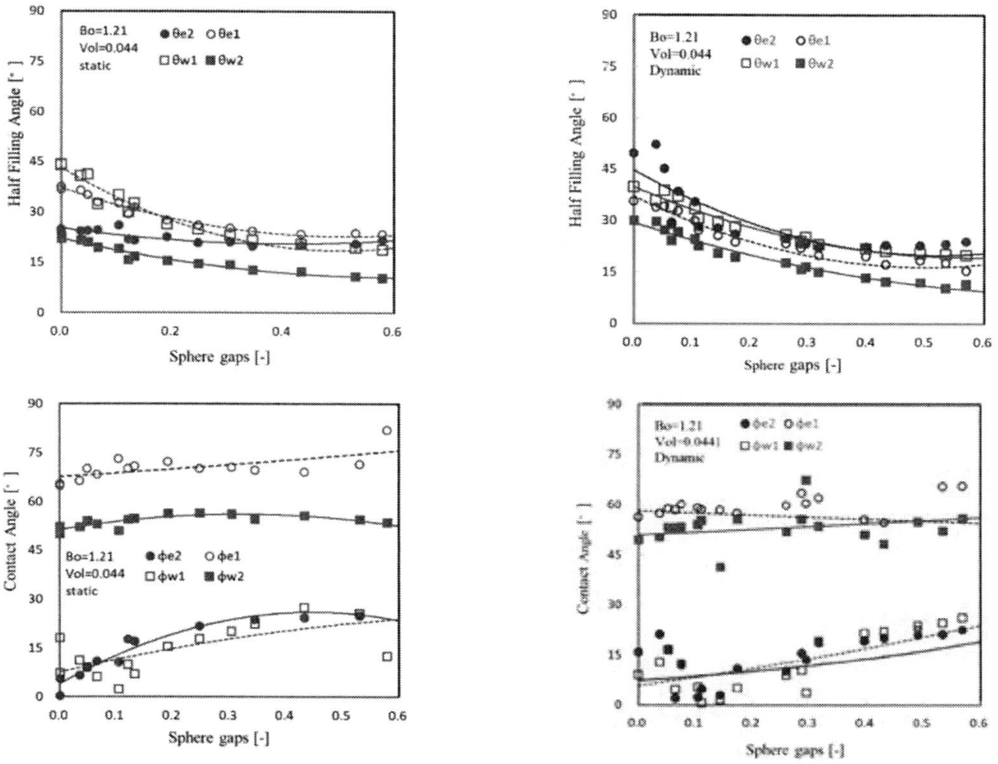

(a) Static state (b) Dynamic state

Figure 10 Profiles of half-filling angles, contact angles with sphere gaps , Bo=1.21,V=0.088

Figures 8 show four types of liquid outlines at static and dynamic state. Magenta and green plots are drawn respectively at the extreme values of half filling angles, contact angles and tensile strengths in Figures 9. During the lower stage displacement two spheres keep at the contact state and the half-filling angles reached to the peak values. Half-filling angles sifted with displacements of a stage have the peak values except the static half-filling angle at the top point of the lower sphere (θ_{Pw2}), which monotonously decrease with stage displacement. The increase and decrease of contact angle respectively corresponds to the advancing and receding contact angle. The contact angles at the bottom point of the lower sphere (θ_{Cw1}) and at the top point of the upper sphere (θ_{Ce2}) transit from the receding to the advancing contact angle due to the infinitesimal values. Figures 10 redraw the half-filling and contact angles with respect to sphere gap. Contact angles contribute to the peak tensile strength, which is measured as resultant force (F_T) of normal (F_n) and tangential (F_t) tensile strength in Figure 11. The normal tensile strength conventionally corresponds to the capillary force. The stage displacement at the peak tensile strength minimize the sum of half-filling and contact angle due to the positive hydrostatic pressure across liquid surface. The larger dynamic tangential strength occurs during the sphere contacting due to the asymmetric deformation. The adhesion force of liquid to sphere at the static state contribute to the larger peak tensile strength.

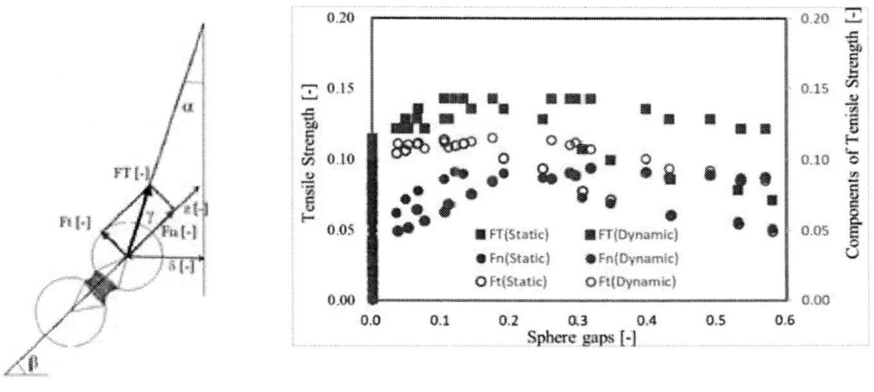

Fig.11 Normal and tangential components of tensile strength at static state

3.4 Numerical simulation at the static state

The solution of Young-Laplace equation demands for one unknown necessary condition on the contact point, i.e. half-filling or contact angle. Liquid surfaces simulated by FEM are drawn with solid lines in Figures 12. The simulation reveals the asymmetric liquid geometry with half-filling and contact angles along longitudinal angle. Disagreements of liquid interfaces results from the approximation of contact lines on spheres. The tangential strength influences depend on the mean curvature. Especially the second figure at the peak half-filling angle has the negative hydrostaic pressure in the direction of relative sphere displacement and positive pressure in the oppsite direction.

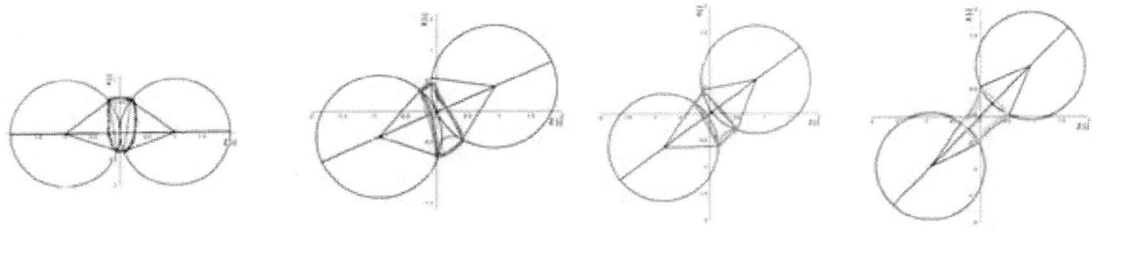

At the initial geometry At the peak halffilling angle At the peak tensile strength At the maximum shpere gap

Figure 12 Numerical estimation at the static state

Figure 13 displays transitions of measured normal tensile strengths ($F_{exp,n}$) and simulated capillary force (F_{cal}) to sphere gap. The initial droplet is attached to the above aperture in the direction of sphere movement due to gravity. The stage displacement drifts parts of liquid volume into the lower aperture of spheres. The sphere on displacement stage keep contacting with the other sphere hang up to the tensiometer up to the peak tensile strength. Fcal and FcalB are numerically estimated by outlines of liquid surfaces respectively with Young-Laplace equation and boundary toroidal model. The sign of hydrostatic pressure with deformation of liquid droplets results in the repulsive capillary force. The normal tensile strength corresponds to the capillary force during the range of symmetric liquid bridge. The angle of beta is defined by the inclination of center distance of a sphere pair. The lower normal tensile strength than the capillary force results from the orthogonal component of capillary force at the low range of beta.

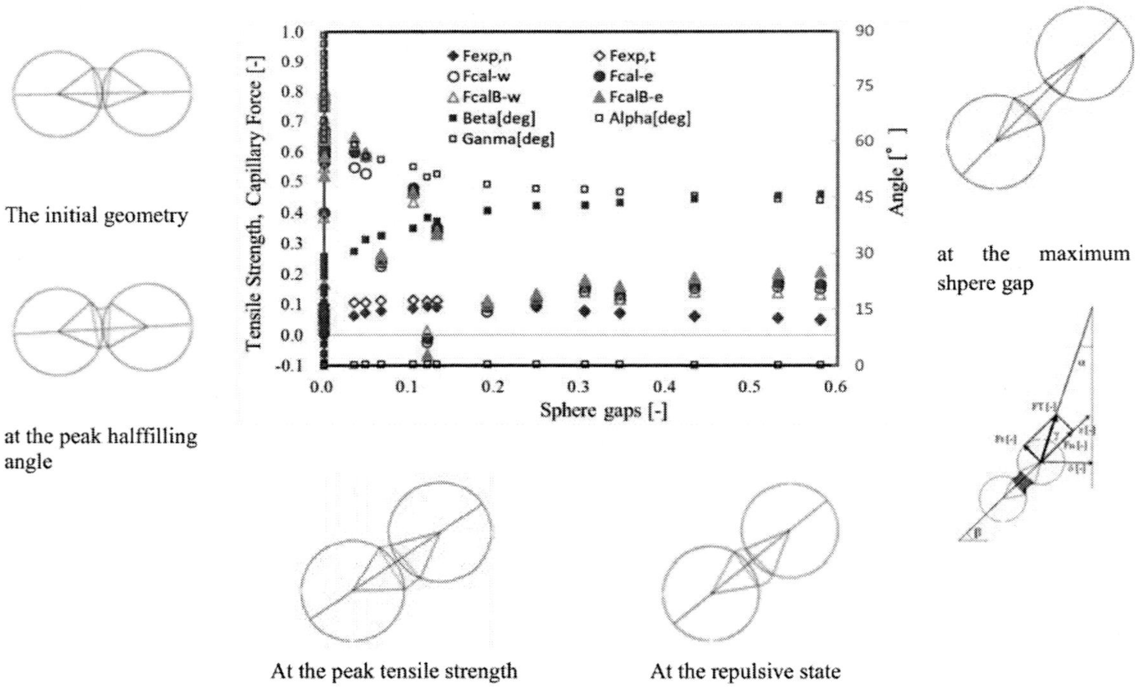

Figure 13 Normal tensile strength and numerical capillary force

Conclusions

Liquid bridges adhered to particles are deformed with the normal and tangential stress in wet granulator. The interparticle force with asymmetric droplet by shear stress must be estimated by the three dimensional solution of Young-Laplace equation. A surface tensiometer and numerical simulation contribute to the normal and tangential tensile strength of liquid bridge adhered to two spheres.

1. The normal strength of symmetric liquid bridge with a surface tensiometer is verified by the conventional toroidal model and the solution of Young-Laplace equation.
2. Sphere gaps at the peak capillary force in the vertical sphere configuration depend on liquid volume and displacement velocity of sphere.
3. The lattice Boltzmann Method is expected to numerically predict the boundary conditions on the contact lines, which are half-filling angle and contact angle at the static and dynamic state.
4. The tensile strength depends on transition of contact angles, which correspond to advancing or receding angles. The larger sum of contact and half-filling angles than the orthogonal angle corresponds to the positive hydrostatic pressure across liquid surface.
5. The normal tensile strength is related with the solution of Young-Laplace equation by FEM. The tangential tensile strength depends on the liquid transition within droplets and the variable mean curvature.

References

[1] K.Murase, H.Shibata, Y.Yasuda and S.Toyama, *Jpn J. Multiphase Flow,* **5**, 322(1991)
[2] A.E.Saez and R.G.Carbonell, *J. Colloid Interface Sci.*, **140**, 408(1990)
[3] D.N.Mazzone, G.I.Tardos and R.Peffer, *Powder Tech.*, **51**, 71(1987)
[4] C.D.Willet, M.J.Adama, S.A.Johnson and J.P.K.Seville, *Langmuir*, **16**, 9405(2000)
[5] M.J.Adama, S.A.Johnson, J.P.K.Seville and C.D.Willet, *Langmuir*, **18**, 6180(2002)
[6] R.K.Niven, *J Petroleum Sci. Eng.*, **52**, 1(2006)
[7] O.Pitois, P.Moucheront and X.Chateau, *Eur. Phys. J.B*, **23**, 79(2001)
[8] S.J.R.Simons, J.P.K. Seville and M.J. Adams, *Chem. Eng. Sci.*, **49**, 2331(1994)
[9] C.D.Willet, M.J.Adama,S.A.Johnson and J.P.K.Seville, , *Powder Tech.*, **130**, 63(2003)
[10] S.M.Iveson, J.A.Beathe and N.W.Page, *Powder Tech.*, **127,** 149(2002)
[11] R.A.Fisher, *J. Agri. Sci.*, 16,492(1926)
[12] F.M.Orr, L.E.Scriven and A.P.Rivas, *J. Fluid Mech.*, **67**, 723(1975)
[13] O.Pitois, P.Moucheront and X.Chateau, *J. Colloid Interface Sci.*, **231**, 26(2000)
[14] U.El Shamy and T.Groeger, *Int. J. Numer. Anal. Meth. Geomeh.*, **32**, 1763(2008)
[15] F.M.Orr, R.A.Brown and L.E.Scriven, *J. Colloid Interface Sci.*, **60**, 137(1977)
[16] K.Murase, T.Mochida, Y.Sagawa and H.Sugama, *Adv. PowderTech.*, **19**, 349(2008)
[17] C.D.Willet, M.J.Adama,S.A.Johnson and J.P.K.Seville, *Handbook of powder technology*, **11**, chap.28, ed. A.D.Salman, M.J.Hounslow and J.P.K.Seville,Elsevier (2007)

NUMERICAL SIMULATION OF WIRE-PLATE ELECTROSTATIC PRECIPITATOR - EFFECT OF PARTICLE CONCENTRATION

Ding Yang[1], *Baoyu Guo*[2*], *Jun Guo*[1], *Aibing Yu*[2]

[1] *Fujian Longking Co. Ltd, Longyan, Fujian, 364000, China*

[2] *Department of Chemical Engineering, Monash University, Clayton VIC 3800, Australia*

Introduction

Electrostatic Precipitators (ESP) have been widely used to separate fine particles from large scale industrial exhaust gases, such as coal fired power plants. In-depth knowledge and careful design optimization are critical to meeting the increasingly strict emission standard in terms of $PM_{2.5}$ emission limit, e.g., 5-10 mg/Nm^3. Numerical modelling has been employed as a powerful tool in this area. Complicated interactions among electric field, gas flow and particle transport are summarized by Adamiak [1]. To model such a system, various levels of simplifications have been made in published papers.

Among them, Lagrangian particle tracking technique is a widely accepted approach. The mass concentration of ash particles in coal-fire flue gas is typically 10-50 g/Nm^3, corresponding to volume fraction in the order of 10^{-5}. Such a small value is often regarded as sufficiently dilute particle concentration in numerical modeling of multiphase flows. For the case of ESP, complicated interactions among the e-field, gas flow and charged particles are involved, even if the deposited dust layer is ignored. Full coupling between the continuous gas phase and discrete particle phase is computationally expensive to achieve a converged steady solution. To avoid such a high cost, the model for ESP was simplified as so called one-way coupling in many previous works, so that the electric field is solved without the particle phase (Guo et al. [2,3]; Farnoosh et al. [4]). This simplification may be acceptable in the end stage of ESP, but will introduce significant error in the early collection stage. Limited studies reported in the literature have considered the particle charges as a source term in the Poisson's equation of electrostatics. For example, Choi and Fletcher [5] predicted a reduced current for the particle-loaded case (d_p = 1-5μm) compared with the particle-free case, but the corona ion charge density somehow shows an increasing trend along the stream. When fine dust particles are treated, the large specific surface area of the particle phase, combined with reduced collection efficiency, implies high space charge density in the collection region. Its effect was understood as a partial suppression of the corona current [6]. As was experimentally proved by Podlinski et al. [7], space charge of charged particles can contribute to the EHD flow pattern and particle transport. In present work, particle space charge is considered through two-way coupling and interesting results are generated in terms of ESP performance as particle loading changes.

Integrated ESP Modelling

Mathematical Equation

Coupled electric field, gas-particle flow and particle charging processes are simulated using

Eulerian-Lagrangian method in Computational Fluid Dynamics (CFD). Major governing equations of vector format are listed below for completeness. For electric field, Poisson equation of potential

$$\nabla^2 V = -(\rho_{ion} + \rho_c)/\varepsilon_0 \quad \mathbf{E} = -\nabla V \tag{1}$$

Current continuity equation

$$\nabla \bullet ((b\mathbf{E} + \mathbf{U}_f)\,\rho_{ion} - D\nabla\rho_{ion}) = S_c \tag{2}$$

where ρ_{ion} is corona charge density, b the ion mobility, D is ion diffusivity, ρ_c is particle space charge density, evaluated by averaging the charges of particles that pass through a control volume (cell),

$$\rho_c = \frac{1}{V_{cell}} \sum_{n=1}^{N} \dot{n}_p q_p \Delta t \tag{3}$$

Where $\dot{n}_p$ is the number flow rate of a parcel of particles tracked, Δt is the residence time in the cell. The source term, S_c, in Equation (2) is transfer rate of ion charges to particles, which is not considered at the moment, i.e., $S_c = 0$. For the gas flow, the momentum equation is,

$$\begin{aligned}\nabla \bullet (\rho_g \mathbf{U}_g \otimes \mathbf{U}_g) = \nabla(p + \tfrac{2}{3}\rho_g k) + \nabla \bullet [(\mu + \mu_t)(\nabla \mathbf{U}_g + (\nabla \mathbf{U}_g)^{\mathrm{T}})] + \rho_{ion}\mathbf{E} \\ -\frac{3\pi n_p d_p \mu}{\mathrm{C_C}}(1 + \mathrm{Re}_p^{0.687})(\mathbf{U}_g - \mathbf{U}_p)\end{aligned} \tag{4}$$

For the particle flow, the force balance of a single particle is

$$m_p \frac{d\mathbf{U}_\mathrm{P}}{dt} = \frac{3\pi n_p d_p \mu}{\mathrm{C_C}}(1 + \mathrm{Re}_p^{0.687})(\mathbf{U}_g - \mathbf{U}_p) + \tfrac{1}{6}\pi d_p^3 \mathbf{g}(\rho_p - \rho_g) + n_e q_e \mathbf{E} + \mathbf{F}_{dis} \tag{5}$$

k-ω SST turbulence model is chosen for gas flow, and the dimensionless Lawless charging rate equation is used for particle charging [8].

Model Conditions

An integrated ESP model developed comprises two parallel modules or submodels, namely, electric field model and coupled gas-particle flow model. One-way coupling between the two modules is used earlier is upgraded to two-way in the present work. The electrostatic field caused by charged particles in suspension is considered in the solution of electric field. The two submodels are coupled through a small under-relaxation factor, in views of strong interaction between the electric field and particles flow, and the stochastic nature of the discrete particle phase. In this way all the volume averaged quantities related to the particle phase (such as the particle space charge density) are smoothed out over time in the simulation.

A single channel, 4-wire and flat plate configuration is considered. The dimensions are 3.5mm in wire diameter, 0.24m for wire-wire spacing, 0.2m for wire-plate spacing. The plate length is 1.2m. The simulation domain is defined as 1.74m long by 0.4m wide box with the 4 holes inside. The grid is smoothly refined at places where higher gradient in electric quantities is expected, such as the proximity of the wires and middle of the adjacent wires. Other base case conditions are as follows.

- Applied voltage on the wire electrode: Va = -60kV. The charge density on the wire surface is determined using an injection law combined with Peek equation (Guo et al.[9]).
- Gas temperature: 120°C. As the inlet boundary condition for the gas phase, the velocity is set as 1 Nm/s; relative turbulence intensity 3.7%, turbulent length scale $0.3D_h = 0.24$m.
- Dust particle size is mono-dispersed at d_p=2.5 μm. Particle density 2000 kg/m^3; ash dielectric constant of κ_p=6.6, Particle loading at the inlet, $(\rho_p)_{in}$, is varied with default value of 10 g/Nm^3.

Typical Results

Electric Field without Particles

When the inlet/outlet boundaries are moved sufficient long distance away from the electrodes in the simulation domain (referred to as "natural boundary condition"), the electric field around the boundary wires show an asymmetric distribution pattern that differ from that around the inner wires. Electric field-lines/current can go out of the limit of the collection region (Figure 1). Electric field intensity on the collection plate is of interest as it is considered a factor to determine the upper limit of operation before spark-over [10]. Normally the field intensity on the plate is highest at the point that faces exactly the wire center with the least distance. This maximum field intensity on the plate is 3.80×10^5 V/m. However, when the natural boundary condition is used, the highest value (1.06×10^6 V/m) predicted in field intensity appears at the endpoint of the plate due to its sharpness. This is a likely point that would trigger electric spark-over prematurely. Practically such a sharp edge should be avoided. Generally, the field strength at the middle wires is smaller than that at a boundary wire, due to the counteraction of adjacent wires. Then the corona charge density at the boundary wires should be greater than that at the inner wires (Figure 2), which is similar to the simulation result of Kasdi [11].

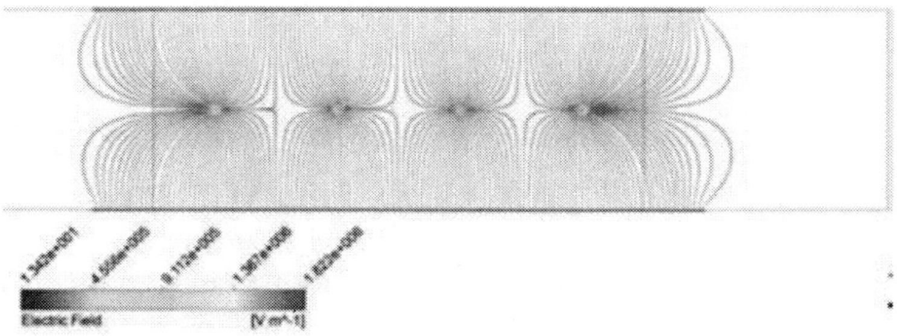

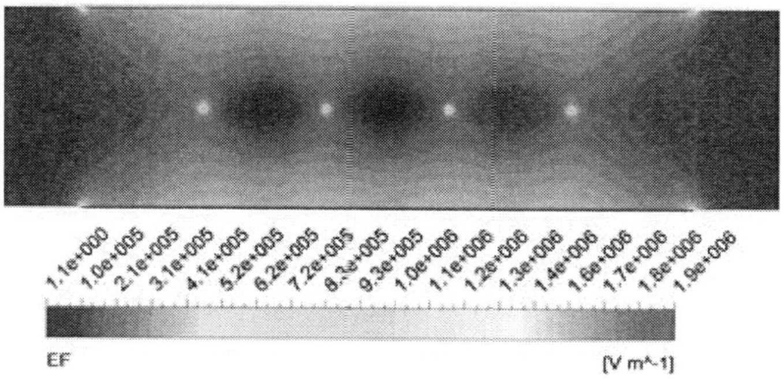

Figure 1. Field streamlines and intensity distribution (no particles).

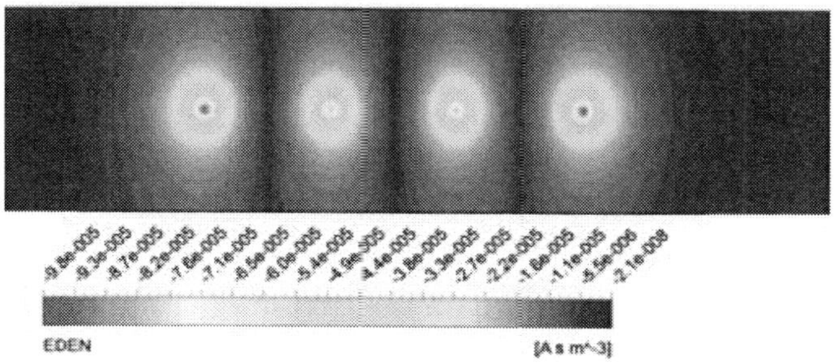

Figure 2. Corona charge density (no particles).

Effect of Particle Space Charge

Particles are randomly injected from the inlet plane with uniform probability in locations and no-slip for velocity (same as the gas velocity). In using random number generator, it is ensured that the injection points are not repeated for every coupling cycle. Figure 3 shows sampled particle trajectories with color scaled to particle charge. Figure 4 shows a pattern of particle deposition rate on the plate. It is seen that the end points of the plate, particularly the downstream one, are special spots to accumulate more dust.

Figure 3. Particle trajectories with colour scaled to particle charge.

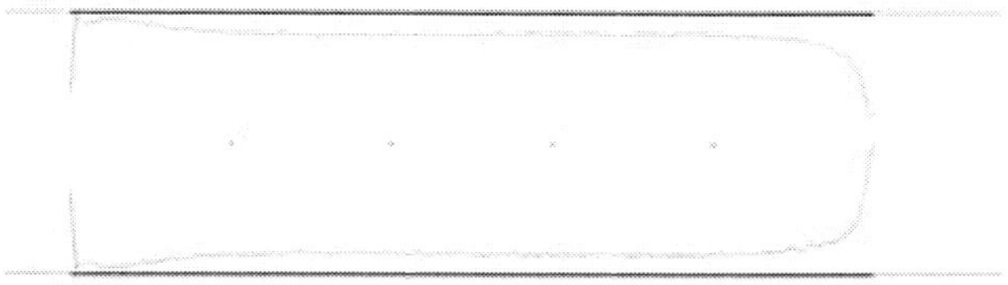

Figure 4. Indicative pattern of particle deposition rate on the plate.

Figure 5 shows the averaged space density of charges carried by particles. Highest space charge density appears near the first wire electrode, where electric current density is highest. It diffuses gradually from the centre line with particles drifting toward the plate while moving forward. Due to the radiating nature of the field-lines around each wire, the lateral component of Coulomb's force in the region close to the centre line may not be as large as that off the centre line. Not collected charged particles of negative polarity are conveyed towards the outlet, lowering the potential downstream outside the collection region (Figure 6), causing distortion of the overall electric field distribution.

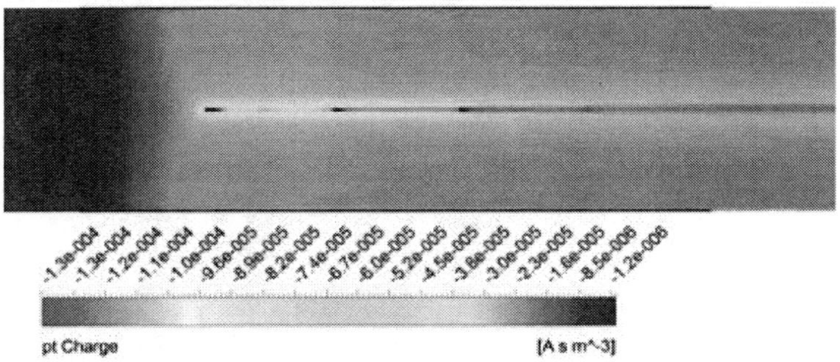

Figure 5. Distribution of space charge density of particles.

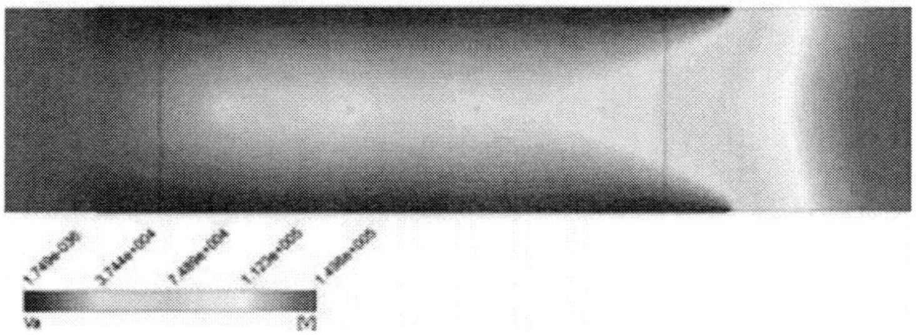

Figure 6. Distribution of electric potential (in absolute value).

Effect of Particle Loading

On Electric Current

Figure 7(a) shows the mean corona current, based on unit length of wire. As can be seen, the corona current drops quickly as particle loading increases, although a fixed voltage is applied. This is caused by the secondary electrostatic field created by the space charges of particles that cancels out the primary electric field at the wire surface and subsequently suppressed the corona generation.

On Particle Collection Efficiency

Predicted particle collection efficiency, as particle loading changes, is given in Figure 7(b). The collection efficiency, increases initially with particle loading (<3 g/m^3), peaks at particle loading of 4-5 g/m^3 and then decreases as particle loading increases further. When ignoring particle space charge, the electric field would be independent of particle loading. This is the case for very low particle loading, where the fractional collection efficiency is constant, or in other words, the quantity of dust collected as well as that penetrated increases proportionally with particle loading.

The observed enhancement of collection by a slight amount of particles is due to the redistribution of electric field so that the near-plate and volume-averaged field intensity happens to increase, which favor the charging of most particles. The asymptotic lower limit of the curve should correspond to the situation when the corona current is so small that turbulent dispersion dominates over the drift force.

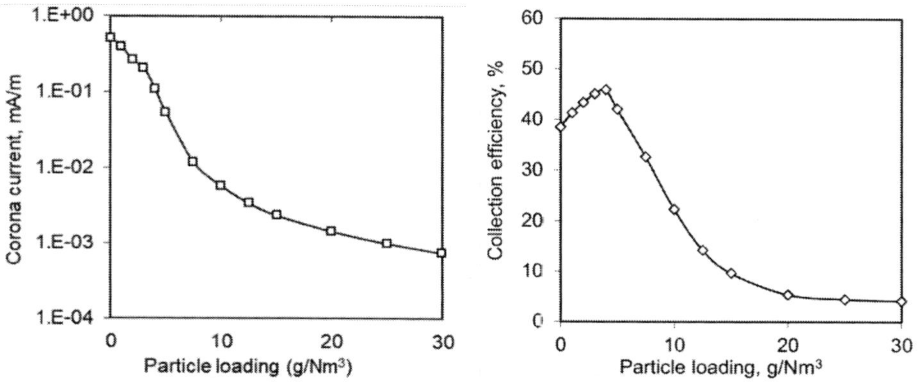

(a) (b)

Figure 7. Mean corona current (a) and collection efficiency (b) as function of particle loading.

On Electric Field Distribution

Figure 8 visualizes the predicted corona charge density and field-lines, which shows how the electric field distribution evolves with particle loading. As the electric field is distorted starting first from the end side, corona density at 3 of the 4 downstream wires is suppressed faster than the 1st wire. When the particle loading is above 7.5 g/m^3, all wires except the first one stop emitting. Meanwhile, reversed field-lines appear around the last wire, and the affected region expands as the loading increases further. Otherwise the field-lines become smoother with increasing loading. Since the field-lines are extended to the outlet, some escaped particles can be pulled back into the collection region, a factor that contributes to significantly greater deposition rate around the end-point of the plate.

Overall, the particle charges affect the particle collection adversely, as they tend to suppress the corona emission or reduce the corona current. Particularly the penetrated particles, which carried significant charges, have prolonged effect as they float along. The corona at the downstream electrodes even extinguish unless the applied voltage is raised individually. When highly charged, more particles precipitate to the plate, reducing the number of particles floating ahead (or less space charge density). The charges carried by these remaining particles exert a negative feedback to the electric field and particle charging, which compromises the collection efficiency.

Electric Quenching

Corona quenching is a common phenomenon in ESP when fine particles are treated. Riebel et al. [12] experimentally studied the corona discharge characteristics of ESP with glycerol aerosol, which found corona onset voltage increases and the corona current significantly decreases after the glycerol steam was injected into the system. Under fixed voltage in the present simulation, corona currents from all the wire electrodes drop with increasing solid loading. The downstream coronas (from the second one) will eventually extinguish at some point. An interesting phenomenon arises when the particle loading increases above 10 g/Nm^3, i.e., the electric streamlines in a small zone around the last wire reverse direction locally. This implies that a charged particle, once falls accidently within the zone, could be trapped and attracted to the same wire, as seen from the particle trajectories in Figure 9. Thus this wire is subject to heavier dust deposition.

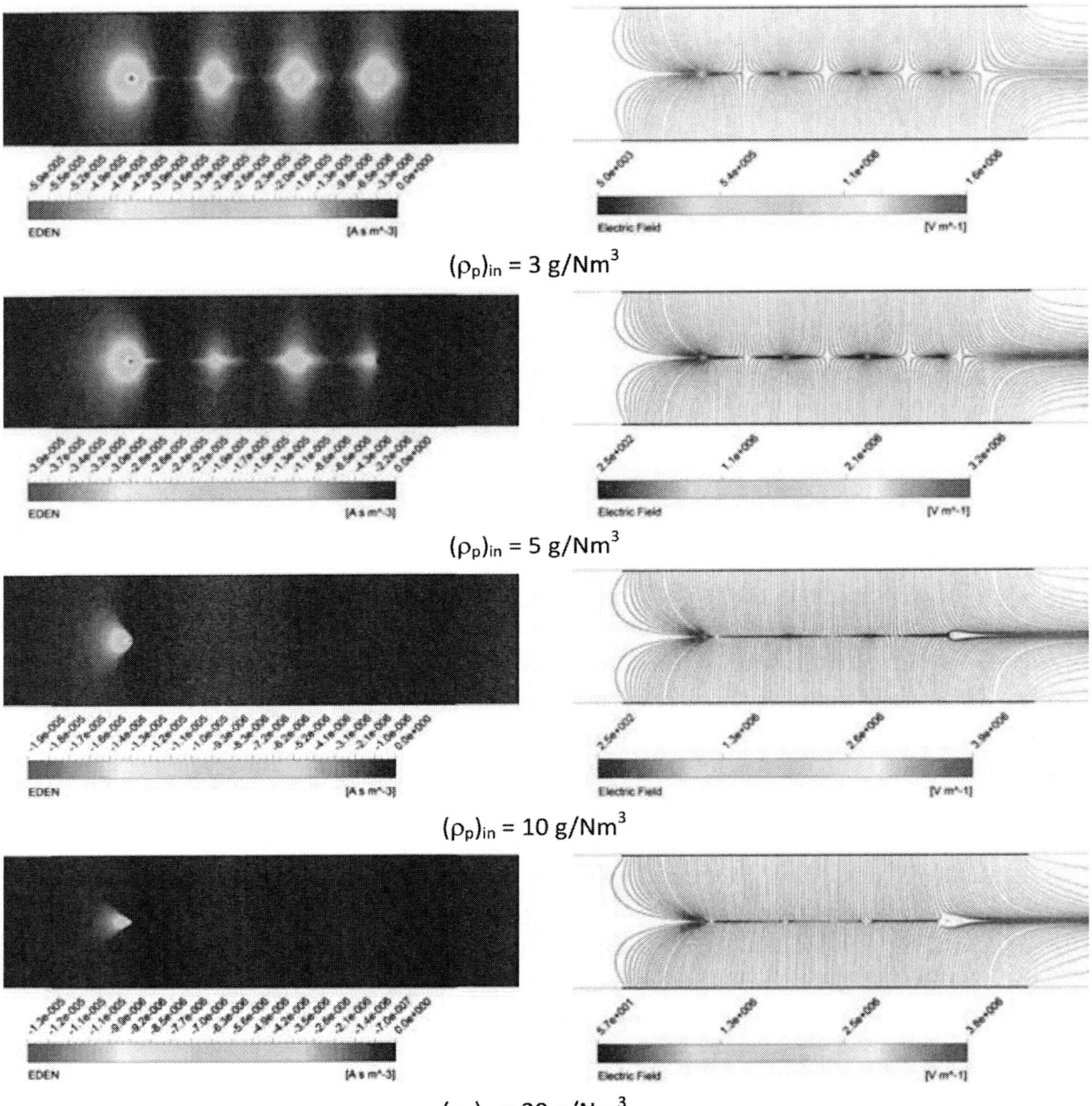

Figure 8. Visualizations of corona charge density and field streamlines for different particle loadings.

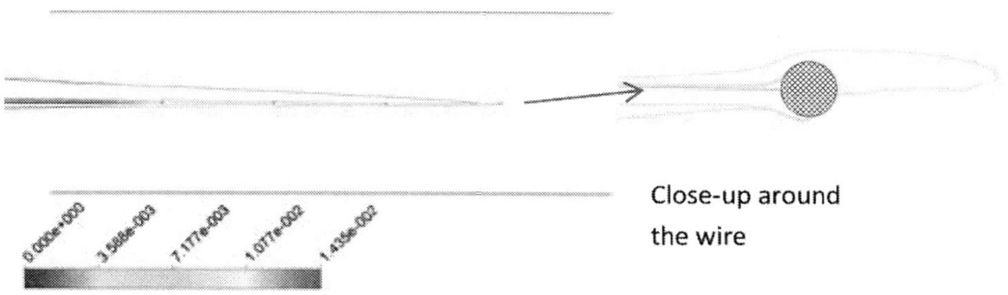

Figure 9. Trajectories of particles falling on the corona wire ($(\rho_p)_{in} = 20$ g/Nm^3).

Concluding Remarks

CFD based ESP model has been advanced, with an enhanced coupling between electric field and particle flow. So the model can be readily applied to the early collection stage of the ESP, where dust particle loading is too high to neglect. Space charge of particles is considered, and its significance is clarified in a broad range of particle mass concentration in the gas. Dust loading significantly affects the dust collection efficiency for fine particles considered. Maximum collection efficiency exists at a critical particle loading that depends on particle size. However, as particle loading increases, collection efficiency decreases and eventually approaches asymptotically to a lower limit. The phenomenon of electrical quenching is observed, where downstream corona is extinguished or one of the emitters is polluted when the particle loading is high.

References

1. Adamiak K. (2013), "Numerical models in simulating wire-plate electrostatic precipitators: A review", *J. Electrostatics*, 71, pp.673-680
2. Guo B.Y., Yu A.B. and Guo J. (2014), "Numerical modelling of electrostatic precipitation: effect of gas temperature", *J. Aerosol Science*, 77, pp.102-115.
3. Guo B.Y., Yu A.B. and Guo J. (2015), "Numerical modelling of ESP for design optimization", *Procedia Engineering*, 102, pp.1366-72.
4. N. Farnoosh, K. Adamiak, G.S.P. Castle (2010), "3-D numerical analysis of EHD turbulent flow and mono-disperse charged particle transport and collection in a wire-plate ESP", *J. Electrostatics*, 68, pp.513-522.
5. Choi B.S. and Fletcher C.A.J. (1998), "Turbulent particle dispersion in an electrostatic precipitator," *Applied Mathematical Modelling*, 22, pp.1009-1021.
6. Parker K.R., Applied Electrostatic Precipitation, Blackie Academic & Professional, 1997.
7. Podlinski J., Niewulis A., Mizeraczyk J. and Atten P. (2008), "ESP performance for various dust densities", *J. Electrostatics*, 66, pp.246-253.
8. Lawless P.A. (1996), "Particle charging bounds, symmetry relations, and an analytic charging rate model for the continuum regime", *J. Aerosol Sci.*, 27, pp.191-215.
9. Guo B.Y., Guo J. and Yu A.B. (2014), "Simulation of the electric field in wire-plate type electrostatic precipitators", J. Electrostatics, 72, pp.301-310
10. White H.J. (1974), "Resistivity problems in electrostatic precipitation", *J. Air Pollut. Control Assoc.* 24, pp.313–338.
11. Kasdi A. (2016), "Computation and measurement of corona current density and V–I characteristics in wires-to-plates electrostatic precipitator", *J. Electrostatics*, 81, 1-8.
12. Riebel U., Radtke R. and Loos R. (2002), "An experimental investigation on corona quenching", *J. Electrostatics*, 54, pp.159-165.

Triboelectric Charge of Spherical Glass Particles against Metal Pipeline

HoSu CHOI [1], Teruo SUZUKI [2] and Kwangseok CHOI [3,]*
[1] Christian Academy in Japan, [2] Kasuga Denki Inc.,
[3] National Institute of Occupational Safety and Health, Japan
**choiks@jniosh.go.jp*

Introduction

Insulating powders are used in many ways such as in the food, pharmacy, and chemical industries. Usually, these powders are highly charged when transported, loaded, dried, and mixed, due to repeated contact and separation between powder particles and the equipment. This results in the powders acquiring electrostatic charges. If not dissipated quickly, electrostatic charge buildup can cause inter-particle cohesion and adhesion to operating equipment, causing a production efficiency decrease. Incendiary electrostatic discharges such as spark discharges, cone discharges, and propagation brush discharges can also take place during industrial operations and may lead to fires and even explosions. Riedewald reported on an industrial fire in a pharmaceutical powder transfer operation [1]. A large number of studies involving electrostatic charge generation on powders, as well as electrostatic discharges and hazards during operation, have been reported. For example, Kaialy reviewed the main factors affecting the electrostatic charging of pharmaceuticals such as particle resistivity, size, shape, and surface roughness [2]. Gibson also reviewed industrial hazards related to electrostatic charges and discharges [3]. Our previous studies [4–6] investigated the electrostatic charging of polymer powder in mixing processes and the ignitability of aluminous, magnesium and polymer powders due to electrostatic discharges involving the effect of the ratio of N2 in the air on the ignitability.

In order to easily measure and confirm the electrostatic charge level of powders, a novel tribocharging system that utilizes the Faraday cup is presented, and the charging behavior of the sample powders is investigated in this paper.

Experimental Apparatus and method

Figure 1 shows the spiral air tribocharge apparatus presented in this study. The apparatus was divided into four main parts: the air supply part, the powder supply part, the tribocharge part, and powder collection part. The air supply part consisted of a compressor (Hitachi, PO3.7MA) that included an auto dryer (CKD, F3-217199), air controller, electronic valve, and switch. In this study, dried air was used because moisture in the air may have an effect on the electrostatic characteristics of the powder sample. The powder supply and tribocharge parts consisted of a powder port, a spiral air flow device, and a pipeline. The spiral air flow device (photo 1) was attached in front of the sample powder port in order to create spiral air inside the pipeline. The spiral air flow device had 8 holes that were 6 mm in diameter and that opened at a slant (60 degrees), which is very different from conventional models [7, 8, 9]. The straight pipeline was made up of stainless steel metal and was 1300 mm in length and 35 mm in diameter. In order to prevent external noise, the overall powder supply and tribocharge parts were shielded. The powder collection part was made up of a cyclone-type Faraday cup, unlike conventional models [7, 8, 9]. The cyclone-type Faraday cup had a separate metal container (30 mm in diameter) for easy measurement of the amount of collected powders. The other devices included two electrometers (Kethley,

6517A), a data logger (Graphtec, GL900), an electrostatic ionizer (Kasuga Denki Inc., KD-510W), an electrical balance (Shimazu, BL220H), and a timer.

The typical operating procedure for the tribocharge test is summarized as follows:
(1) Adjust the air pressure using the air pressure valve.
(2) Load 1 g of the sample powder into the powder port.
(3) Eliminate the initial charge of the sample powders using an electrostatic ionizer.
(4) Reset the electrometers (Q mode) connected to the pipeline and cyclone.
(5) Switch the air flow on, and run the sample powder through the pipeline where friction occurs.
(6) Measure the indication value when it stabilizes.
(7) Measure the amount of powder collected in the Faraday cup.
(8) Clean out the pipeline and cyclone part using an air spray in preparation for the next test.

The charge to mass ratio, q (nC/g), of the sample powder was obtained by dividing the charge, Q (nC), by mass, m (g), of the sample powder loaded into the Faraday cup:

$$q = \frac{Q}{m} \quad \text{(1)}$$

This experiment is mainly to determine how much the charge to mass ratio, q (nC/g), of the sample powders is affected by the size of the powder and the amount of air pressure. The cyclone-type Faraday cup and the pipeline were suspended electrically with insulators. All experiments were conducted in ambient conditions (air temperature: 22 ℃; relative humidity: 55 %).

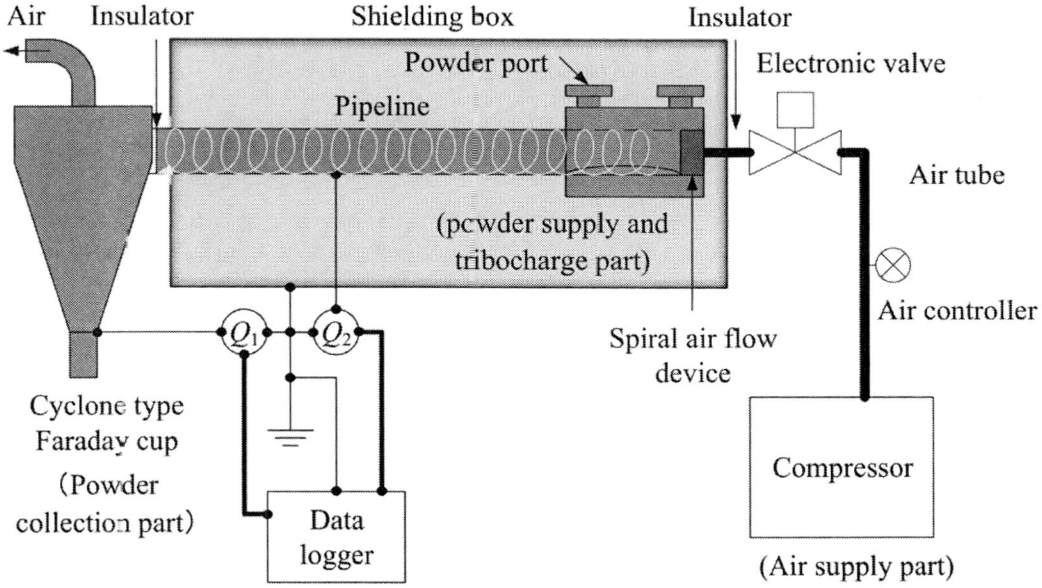

Fig. 1 Schematic of the spiral air tribocharge apparatus used in this study.

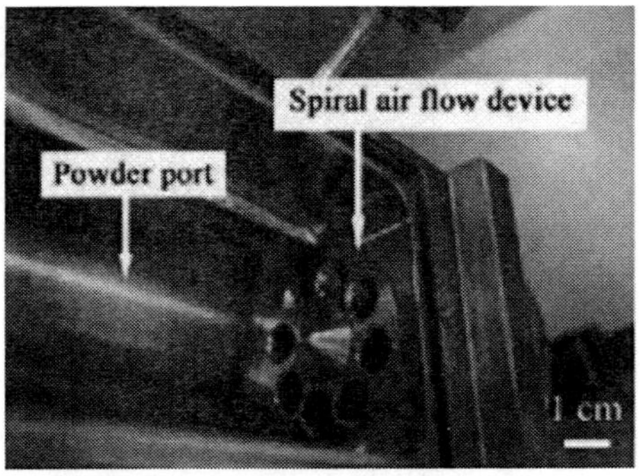

Photo 1 Structure of the powder port and spiral air flow device.

Sample powders

Glass beads are extremely good light reflectors and thus are widely used as the main component of the coating material for light-reflecting adhesive tape, background paint for traffic control sign boards, and paint for road signs. Four different grades of commercial glass beads were used in this experiment, as shown in Table 1. The particle size was determined using a laser diffraction system (LDA Win 1.21, dry type). Hereafter, they will be called by their abbreviation shown in Table 1. The analysis of the images from microscopy (Neoscope, JCM-500) revealed that the G-4 powder particles were spherical, as shown in photo 2. The other glass bead particles were the same shape as G-4.

Table 1. List of powder specimens used in experiments.

Specimen	Abbreviation	Diameter [μm]		
		D_{10}	D_{50}	D_{90}
Glass beads 1	G-1	22	62	85
Glass beads 2	G-2	93	131	175
Glass beads 3	G-3	130	180	241
Glass beads 4	G-4	162	225	294

D_X: X% volume average.

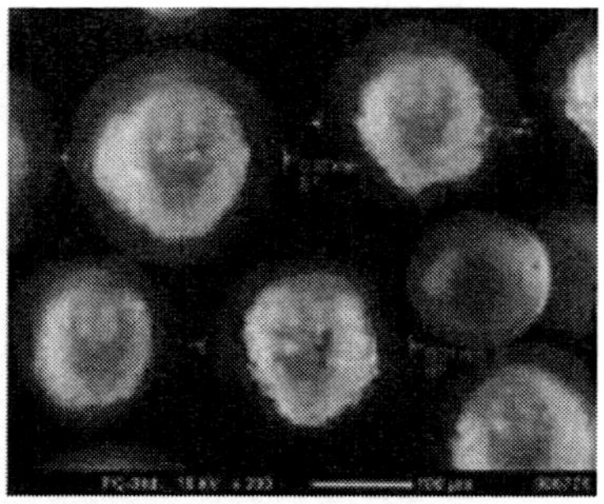

Photo 2 Shape of G-4 powder used in this experiment.

Experimental results and discussion

Figure 2 shows the typical change of charge detected over time when the G-3 particle was used and the air pressure was 0.15 MPa. It should be noted here that when no particle was used and only air was passed through the pipeline, the charge was detected as almost zero. The charge Q1 of the powder was found through the cyclone-type Faraday cup and then was confirmed through the charge Q2 obtained from the pipeline. As the powder particles passed through the pipeline for about 0.82 seconds, as seen in the Fig. 2, the charge clearly increased. Q1 and Q2 had almost the same trend and absolute values. This result confirmed the reliability of the test and showed that there was no disturbance by external noise.

In this study, all of the glass beads (G-1 to G-4) were negatively charged when stainless steel metal was used as the contact surface. The charge polarity of the materials relates to the work function (or ionization potential) of each material. The contact potential difference between two substances is defined by the difference between their fermi levels, which correlate with the work function. In general, materials with higher work functions tend to appropriate electrons from materials with lower work functions. This point will be carried out in our further study.

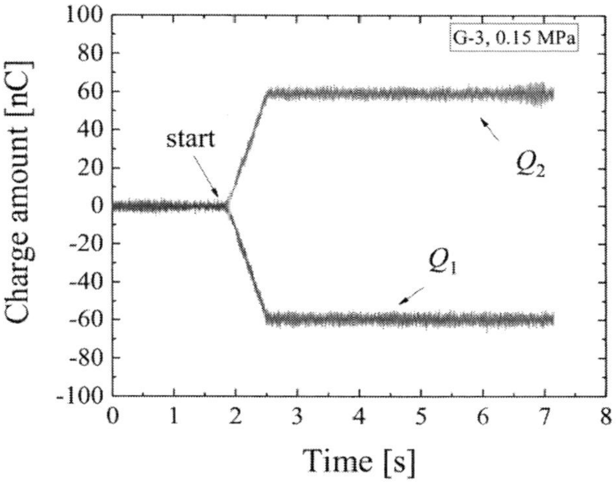

Fig. 2 Change of charge detected over time.

Figure 3 shows the charge to mass ratio of glass beads as a function of applied air. The range of applied air was between 0.1 MPa and 0.3 MPa in this study. When the applied air pressure was below 0.1 MPa, it was most likely not enough to send all of the powder through the pipeline. Each experiment was repeated five times, and the average values were reported. This graph does not contain an error bar, but these were about 10%. The results showed that the charge to mass ratio of both powders increased as the applied air increased. Although not shown in Figs. 2 and 3, the increase in air pressure also causes the slope of the graph to increase significantly. This can be explained by the fact that with increased pressure of applied air, the particles collide against the surface of the pipeline with more strength [9]. In order to investigate in more detail, the velocity of air at approximately 10 mm from the inside of the pipeline wall was measured using an anemometer (Kanomax, Anemomaster-6115). The approximate velocity of air flow was 9.5 m/s at 0.1 MPa, 12 m/s at 0.15 MPa, 15 m/s at 0.2 MPa, 20 m/s at 0.25 MPa, and 22 m/s at 0.3 MPa. It could be clearly seen that as the air pressure increased, the air velocity also increased. Another reason for this may be that because the air was applied in a spiral form, the particles may have collided more frequently against the pipeline when the air flow was increased. This exception will be investigated in detail in the future.

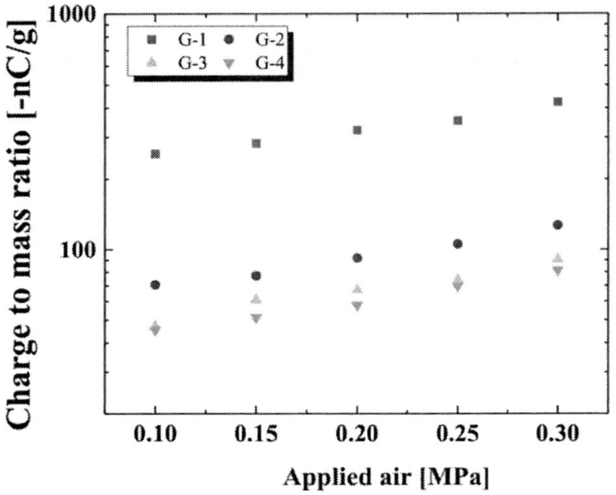

Fig. 3 Charge to mass ratio of glass beads as a function of applied air.

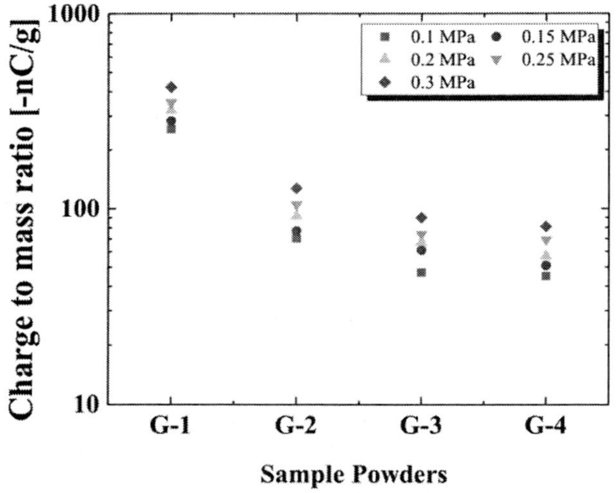

Fig. 4 Charge to mass ratio as a function of different glass bead sizes.

Figure 4 shows the charge to mass ratio as a function of different sizes of glass beads. The size distribution of each glass bead in this study was described in Table 1. The results showed that the charge to mass ratio of the glass beads increased as the powder size decreased. That is to say, smaller particles had a larger q than the larger ones. The total surface area of all of the smaller beads combined is greater than that of the larger ones, which results in a high charge to mass ratio development in the tribocharge. The value range of the q of the glass beads obtained in this study was about 40 nC/g to 100 nC/g. These values agreed with those obtained from the micronizing and pneumatic transfer process [10]. These values were high enough to generate incendiary electrostatic discharges during handling in industrial processes. This suggests that if glass beads are used with solvents in the cleaning process, they must be handled carefully, as incendiary discharges can ignite the solvents even though the glass beads themselves are nonflammable material.

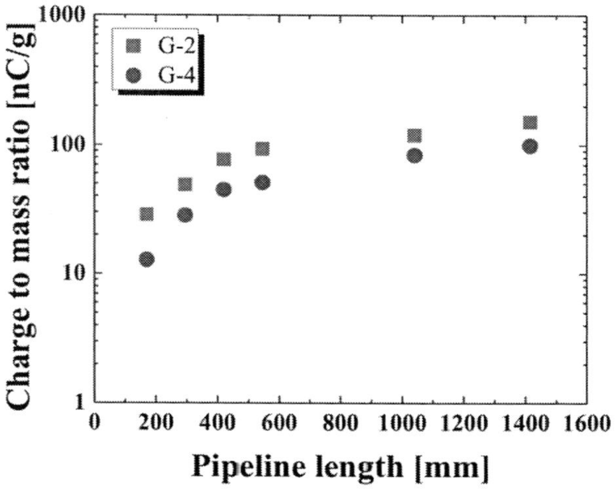

Fig. 5 Charge to mass ratio as a function of pipeline length.

Figure 5 shows the charge to mass ratio as a function of pipeline length. Six different pipeline lengths, *d*, were used: 170, 295, 420, 545, 1040, and 1415 mm. The glass beads used in this experiment were G-2 and G-4, and the air pressure was 0.35 MPa. Since the pipeline length changed with each test, it was difficult to attach the Faraday cup to the end of the pipeline. Therefore, in this study, only the value of q_2 was measured. As seen in Fig. 2, q_2 had the almost same value as q_1 but had an opposite sign. The length of the pipeline was measured starting from beginning of the powder port because the inside structure of the powder port was the same as that of the pipeline. Each experiment was repeated five times, and the average values were reported. The results showed that q_2 increased as the length of the pipeline increased for both sizes of glass beads. This can be explained by the fact that as the length increases, the time the glass beads come in contact with the pipeline also increases. The increase rate at the beginning section of the pipeline was especially large. It can also be seen through Fig. 5 that overall, G-2 had a larger q_2 than G-4 because of the total surface area. This confirms the reliability of the data shown in Fig. 4.

In addition, both beads seemed to reach saturation when the pipeline length was 1415 mm. It was compared to the maximum charge to mass ratio, q_c [nC/g] calculated from the well-known equation shown below [11]:

$$q_c = |\sigma_{max}|/(D \cdot \rho), \quad \text{-------------------------}(2)$$

where σ_{max} [kg/m^3] is the maximum surface charge density (27 μC/m^2); D [μm] is the particle diameter; and ρ is the particle density (1.5×10^3 kg/m^3).

In this study, the particle sizes of G-2 and G-4 were 131 μm and 225 μm, respectively, as shown in Table 1. The calculated maximum values q_c and experimental maximum values q_t are compared as shown in Table 2. The calculated and experimental values for both G-2 and G-4 agreed well with each other. Therefore, it can be seen that both G-2 and G-4 reached saturation.

Table 2. Calculated maximum values and experimental maximum values of the charge to mass ratio.

	q_c [nC/g]	q_t [nC/g]
G-2	137	151
G-4	80	99

Conclusion

In this paper, a novel tribocharging system that utilizes the Faraday cup was presented and the electrostatic charge behavior of glass beads was experimentally investigated. The results were as follows:
1) For all of the glass beads, the charge clearly increased as the glass bead particles passed through the pipeline. The charge amount of the glass beads and the charge amount obtained from the pipeline had almost the same trend and absolute values. This result confirmed the reliability of the test and showed that there was no disturbance by external noise.
2) All of the glass beads were negatively charged when stainless steel metal was used as the contact surface. The value range of the charge to mass ratio of the glass beads obtained in this study was about 40 nC/g to 100 nC/g.
3) The charge amount of the glass beads increased as the applied air increased. It could also be seen that as the air pressure increased, the air velocity also increased.
4) The charge amount of the glass beads increased as the powder size decreased due to the total surface area of the beads.
5) The charge amount of the glass beads (G-2, G-4) increased as the length of the pipeline. The increase rate at the beginning section of the pipeline was especially large.

Acknowledgments
The authors gratefully acknowledge Mr. Naoto Nogera for his assistance during the test.

References

1. F. Riedewald (2012): "A fire in a secondary pharmaceutical powder transfer operation," *Process safety progress*, 31, 4, pp. 390-392.
2. W. Kaialy W, "A review of factors affecting electrostatic charging of pharmaceuticals and adhesive mixtures for inhalation," *Int.J.Pharm*. Vol.503, pp.262-276.
3. N. Gibson (1997), "Static electricity — an industrial hazard under control," *J. Electrostatics,* Vol. 40-41, pp. 21-30
4. K. Choi, K. Choi, K. Nishimura (2015), "Experimental study on the influence of the nitrogen concentration in the air on the minimum ignition energies of combustible powders due to electrostatic discharges," *J. of Loss Prevention in the Process Industries.* Vol.34.pp.163-166.
5. K.S. Choi, H.Sakasai, and K. Nishimura (2015), "Experimental study on ignitability of pure aluminum powders due to electrostatic discharges and Nitrogen's effect," *J. of Loss Prevention in the Process Industries.* Vol.35.pp.232-235.

6. K.S. Choi, H. Sakasai, and K. Nishimura (2015), "Minimum ignition energies of pure magnesium powders due to electrostatic discharges and Nitrogen's effect," *J. of Loss Prevention in the Process Industries.* Vol.41.pp.144-146.
7. N. Masui, and Y. Murata (1981), "charging tendency of the flowing powder through a metal tube (II), " *The five Annual meeting of Institute of Electrostatic Japan,* pp. 244-247.
8. S. Watano, S. Saito, and T. Suzuki (2002), "Feedback control of powder electrostatic charge in a pneumatic conveying process," *J. Soc. Powder,Japan*, Vol. 39, pp. 496-502.
9. P. Boschung, M. Glor (1980), "Methods for investigating the electrostatic behavior of powders, J. of elelctrostatic," Vol. 8, pp. 205-219.
10. IEC50404(2004), "Electrostatics- Code of practice for the avoidance of hazards due to static electricity," p. 66.
11. M.Glor (1988), "Electrostatic hazards in powder handling," *Research studies press LTD.*, p.26.

509266 Spray-Drying of a Layered Double Hydroxide Nanosuspension

Abstract:

Over the past few years, layered double hydroxide (LDH) has attracted particular interests in both academia and industry. LDH is 2D anionic clay of brucite-like layered structure with various organic or inorganic anions located between layers. The diverse composition of LDH, and its associated physical and chemical properties open up new possibilities in various applications. LDHs are successfully applied as precursors in synthesis of supported catalysts, sorbents, carriers delivering bioactive molecules, matrices containing active pharmaceutical ingredients for control release, electrochemical biosensors, etc. LDHs are mainly synthesized in a form of nanosuspensions. The application of nano-sized sorbents or catalysts in the liquid-phase processes requires a special technique to separate them from the final product. An uncontrollable agglomeration of solid nanoparticles makes their industrial application impossible in a dry phase. Moreover, the toxic effects of nanoparticles on human health and ecosystem have been discussed in the literature.

Utilization of larger micron-size agglomerates of nanoparticles is one of the alternative ways to safely handle nanoparticles in the dry phase on the industrial scale. Spray drying is one of the well-known techniques that have been widely applied for production of microspheres from nano-sized particles. Using this technique, it is possible to control the size distributions as well as the morphology of spray-dried agglomerates. However, there are only a limited number of publications related to spray drying of LDH nanosuspensions. Moreover, spray-dried LDH agglomerates of rather wide size distribution, low surface area and irregular shape were reported in the literature.

In this study, we explore the effect of process parameters, such as spraying air flow rate, drying air temperature, slurry concentration and slurry feed rate on the size distribution, morphology, specific surface area and pore volume of LDH agglomerates fabricated by spray drying of nanosuspensions.

The round-shaped agglomerates with a diameter of more than 2 and less than 15 micrometers were obtained by spray drying of suspensions of nanoparticles with median diameter of 171 nm without addition of dispersant and binder. Our results confirmed that the spraying air flow rate has the dominant effect on the agglomerate median size and the width of agglomerate size distribution as well as on the agglomerate specific surface area. The small agglomerates with narrow size distribution were collected during drying at the high flow rate of spraying air. Our results also indicate that lowering the temperature of drying air and using the dilute nanosuspensions promote generation of small agglomerates.

The specific surface area of spray-dried LDH agglomerates is in the range of 49 to 101 m^2/g. The agglomerates of high specific surface area were obtained at low flow rate of spraying air, high slurry feed rate of concentrated slurry and using low temperature drying air. These results also confirm that drying at low drying rate yields to formation of agglomerates of high surface area. Similar trends were observed in terms of the variation of total pore volume with drying process parameters.

Thus, by adjusting the spray drying conditions, it is possible to produce agglomerates of LDH nanoparticles satisfying the product requirements.

513829 Impact of Spheroidization of UO_2 Powders on the Filling of Press Molds

Abstract:

In industry, the use of powders is very common. This underlines the interest in understanding the flow of powders in process lines (hopper, silos, transportâ€¦) and finding the link between powder physical characteristics and its rheological properties. This study deals with the flowing properties of a UO2 powder in a powder metallurgy process for nuclear fuel fabrication. A reproducible filling of the press molds is required to avoid mechanical grinding of the sintered pellets.

A powder is characterized by its particle size distribution but also by the morphology of particles. Morphology has a major impact on the cohesion and then on the flowability of powders [1]. UO2 powders are generally made up agglomerates which are composed of aggregates, themselves formed by crystallites (Figure 1). Agglomerates are primarily governing the powder flowing behavior. Their characteristics sometimes do not allow them to meet the expectations of the process, especially for the filling of the press molds.

For this reason, in a powder manufacturing process, a spheroidization operation (Figure 2) could precede the shaping of pellets to improve the filling properties.

In this study we observe the influence of particles morphology on the rheological properties of UO2 powders. Whereas there is no standard and universal method to characterize the flow of powders [2], there is direct (shear cells) and indirect methods (dynamic flowability measurements like permeability). We use both methods: the indirect methods based on qualitative measures of powder cohesion, the direct methods to determine the powder cohesion for a specific consolidation state through the Freeman shear cell [3]. While a shear cell measurement permits to determine the powder flow in quasi-static system, a permeability measurement shows the flow index in dynamic system. We identify the link between theses measurements and the quality of the filling.

References:
[1] Xiaowei Fu and others, «Effect of Particle Shape and Size on Flow Properties of Lactose Powders», Particuology, Advances in Characterization and Modeling of Particulate Processes, Apr. 2012.
[2] P. Juliano, B. Muhunthan, and G. V. Barbosa-Cánovas, « Flow and shear descriptors of preconsolidated food powders », J. Food Eng., vol. 72, no 2, p. 157-166, Jan. 2006.
[3] J. Schwedes and D. Schulze, « Measurement of flow properties of bulk solids », Powder Technol., vol. 61, no 1, p. 59-68, Apr. 1990.

508203 Modelling a Twin Screw Granulator Using the Discrete Element Method

Abstract:

Wet granulation is a process used to create larger stable agglomerates (granules) from fine powders. This has many desirable outcomes such as improving flowability, compactibility, and homogeneity. Granulation is commonly employed in the food, pharmaceutical, detergent, and fertilizer industries, but despite its wide adoption it is often inefficiently operated, with high recycle ratios in continuous processes and high rejection rates in batch processes. Wet granulation is the most common type of granulation and, in pharmaceutical applications, it is a critical step in tablet manufacturing that affects the uniformity and compactibility of the final dosage form.

Traditionally batch granulation was the favoured granulation process method in the pharmaceutical industry due to the challenges faced from adopting a continuous approach, such as cost or inability to monitor product quality reliably. Since the introduction of Quality by Design (QbD) and Process Analytical Technology (PAT) by the FDA in 2003 there has been a move towards new manufacturing techniques such as continuous processing.

Although tremendous efforts have been made to gain scientific insight into the granulating process, a fundamental understanding of wet granulation is still lacking due to the complexity of the mechanisms involved. With twin screw granulation becoming a popularly employed method of wet granulation, an in-depth understanding of particle enlargement in the granulating process is necessary in order to improve the quality of the final product without the need for large-scale Design-of-Experiment studies.

Despite the extensive experimental research carried out for twin screw granulators in recent years, there has been little computational work carried out. This paper employs the Discrete Element Method (DEM) to study a 25 mm diameter, GEA ConsiGmaâ„¢ 1 twin screw granulator with a typical 60Â° forward configuration for kneading elements. The DEM simulations were conducted using the commercial code EDEM with a DEM contact model developed for cohesive solids implemented through an API. The contact model is based on an elasto-plastic contact with adhesion and uses hysteretic loading and unloading paths to model the elastic-plastic contact deformation. In these simulations, the adhesion is used to capture the effect of the binder liquid without the complication of modelling the liquid directly. The adhesion parameter is a function of the plastic contact overlap. The model has previously been shown to be able to predict the stress-history-dependent behaviour depicted by a flow function of the material. In this study the effect of particle size, gap tolerance and cohesion on the residence time distribution and local solid fraction at different elements in the granulator is assessed. The level of cohesion used is related back to uniaxial test results which can be used to measure the flowability of a cohesive granular solid.

508216 Hybrid Multiscale Modelling of a Twin Screw Granulator

Abstract:

Wet granulation is a process used to create larger stable agglomerates (granules) from fine powders. This has many desirable outcomes such as improving flowability, compactibility, and homogeneity. Granulation is commonly employed in the food, pharmaceutical, detergent, and fertilizer industries, but despite its wide adoption it is often inefficiently operated, with high recycle ratios in continuous processes and high rejection rates in batch processes. Wet granulation is the most common type of granulation and, in pharmaceutical applications, it is a critical step in tablet manufacturing that affects the uniformity and compactibility of the final dosage form.

Modelling processes such as wet granulation is not straightforward, but may be achieved using techniques such as Population Balance Modelling (PBM) and Discrete Element Method (DEM). PBM groups particles into different classes based on their properties and uses empirically derived differential equations to describe key processes such as aggregation or breakage. However, although PBM tracks the size and number of particles over time, it cannot account for the distribution of the particles and the effect of this local variability.

By contrast, DEM simulations track individual particles, and as such, phenomena such as segregation or liquid migration can be studied. DEM simulations also consider the effect that the geometry and its dynamics have on the particles' behaviour, which is something that PBM cannot account for directly. However, DEM simulations of fine materials can be computationally expensive and this often results in reduced numbers of particles being used to make the simulation computationally tractable. This is usually achieved through increased, unrealistic particle sizes or the simulation of isolated local zones without considering adjacent zones which may affect the behaviour.

Traditionally, large-scale Design-of-Experiment (DoE) studies would be needed to estimate key rate parameters for use in PBM modelling of granulation and this can be both costly and time-consuming. DEM simulations provide an alternative method to estimate these parameters. By coupling the two simulation methods, one can utilise the key benefits of each method to better simulate complex processes.

This paper presents a hybrid approach to modelling twin screw granulation via DEM–PBM coupled simulations to study a 25 mm diameter, GEA ConsiGma™ 1 twin screw granulator with a typical 60° forward configuration for kneading elements. The DEM simulation was conducted using the commercial code EDEM with a DEM contact model developed for cohesive solids implemented through an API. The PBM simulation is carried out with the commercial code gFORMULATE which is built on top on the advanced modelling platform gPROMS. A bi-directional coupling is implemented with information shared between the two simulation tools at specified intervals. Preliminary results for this coupled approach are presented and discussed.

509577 Blocking Rules for Discharging Granular Materials from a Flat Bottom Hopper

Abstract:

Hoppers and silos are widely used in storing granular materials in various industries, such as agricultural, chemical, food and pharmaceutical industries. Three key aspects need to be considered when hoppers are used to store powders and grains: 1) stress analysis; 2) hopper blockage ; 3) mass flow rate. A hopper must be designed strong enough to sustain the stress generated from the bulk materials during storage and discharge. On the other hand, a hopper must be designed properly so that the stored bulk materials can be discharged, i.e. without blockage and formation of ratholes and arches. Furthermore, the mass flow rate needs to be predicted and controlled. In this study, a combined experimental and numerical study was performed to determine the blocking rule for discharging granular materials from a flat bottom hopper. Granular materials of various sizes were considered and the minimal orifice size through which the bulk materials can be discharged was determined experimentally using a Flodex tester and numerically using DEM. The dependency of the minimal

orifice size on particle size was explored. An semi-empirical model was also developed. Using this model, the critical orifice size at which the flow will stop, i.e. blockage takes place, can be determined from the mass flow rate data. It was found that excellent agreement between the numerical results and experimental measurements was obtained. Both experimental and numerical results showed that, even for spherical particles, layered dead zones could be formed during the discharge from the flat bottom hopper. In addition, mono-layered dead zones were developed when large particles were used. Furthermore, the developed blocking rules were also verified experimentally and numerically.

508310 Characterization of Mesoscopic Structure in Cohesive Powder, Neat or Blended, By X-Ray Computed Tomography and Prediction By the Discrete Element Method

Abstract:

Understanding and predicting structural characteristics of cohesive powder systems is a technical challenge in many powder technology processes and applications. The discrete element method (DEM) [1] is a promising approach to model the structure and flow behavior of cohesive granular systems. Advances in experimental techniques such as X-ray computed tomography (XRCT) have now enabled characterization of fine cohesive powder systems with sub-micron resolution. Combined with standard bulk powder characterization methods (e.g. density), this technique allows for mesoscopic structural characteristics of cohesive powder systems to be compared with DEM simulation. In this study, DEM and XRCT are used to characterize the mesoscale structure, under controlled consolidation conditions, of two moderately cohesive 'model' (roughly spherical) powders with distinct material properties. The DEM simulations are based on a constitutive elasto-plastic model [2] that makes use of the JKR theory [3] to account for cohesive interparticle forces. Density and void structure obtained from DEM simulation are compared to characterization results obtained from XRCT image analysis. XRCT and DEM simulation approach is applied to examine structures of binary mixtures of the two model powders. The ability of DEM simulation and the contact model to accurately predict structural characteristics of single component and binary granular systems is discussed.

References

[1] P. A. Cundall and Otto D.L. Strack. A discrete numerical model for granular assemblies. Geotechnique, 29(1):47–65, 1979.

[2] C. Thornton and Z. Ning. A theoretical model for the stick/bounce behavior of adhesive, elastic-plastic spheres. Powder technology, 99(2):154– 162, 1998.

[3] K.L. Johnson, K. Kendall, and A.D. Roberts. Surface energy and the contact of elastic solids. In Proceedings of the Royal Society of London A: Mathematical, Physical and Engineering Sciences, volume 324, pages 301–313. The Royal Society, 1971.

511638 Numerical Research of Hydrodynamics in Gas-Solid Micro Fluidized Beds

Abstract:

Due to the advantages of less energy consumption, more safety and good operability, the miniaturization of fluidized bed, or named micro fluidized bed, has been extensively studied by a lot of researchers. The Micro Fluidized Bed Reaction Analyzer (MFBRA) was adopted by the Institute of Process Engineering (IPE), Chinese Academy of Sciences (CAS), to study the gas-solid reaction kinetics under isothermal differential condition[1, 2]. In the MFBRA, the hydrodynamics of micro fluidized beds play a vital role on the heat and mass transfer and which is a key to the accurate measurement of gas-solid reaction kinetics. In order to optimize and design the MFBRA, numerous experiments have been conducted to study the hydrodynamics of micro fluidized beds[3, 4]. Unfortunately, those works mostly focused on the bed pressure drop and minimum fluidization velocity, nevertheless, the detail information, such as local solid volume fraction and particle velocity profile in micro fluidized beds are reported rarely. It is because the existing experimental techniques, such as, optical fiber and capacitance probes are hard to measure those local parameters without disturbing the flow field.

The objective of this work is to study the hydrodynamics of micro fluidized beds and to investigate the isothermal characteristics during kinetics testing of graphite powder combustion in MFBRA. In this study, Discrete Particle Model (DPM) and Two-Fluid Model (TFM) are selected to investigate micro fluidized beds, respectively. The DPM is carried out by the commercial software EDEM-Fluent Coupling simulations. And for TFM, we used the Eulerian multiphase model in the ANSYS Fluent. From our numerical simulations, the detailed hydrodynamic characteristics of micro fluidized beds are explored systematically by comparing different column sizes and superficial gas velocity. Meanwhile, it is found that both DPM and TFM have the ability to model the MFBRA under practical operation condition with high superficial gas velocity. In the term of isothermal characteristics in micro fluidized beds, the temperature field distribution as well as the heat transfer between heat-carrier and graphite powder are explored by the combination of simulations and experiments.

Reference:

1. Xu, G., *Micro fluidized bed reaction kinetics analyzer. Supported by instru-ment development project of Chinese Academy of Sciences, China, Project numberY2005014. (in Chinese)*. 2005.
2. Yu, J., et al., *Isothermal differential characteristics of gasâ€"solid reaction in micro-fluidized bed reactor.* Fuel, 2013. **103**: p. 29-36.
3. Liu, X., G. Xu, and S. Gao, *Micro fluidized beds: Wall effect and operability.* Chemical Engineering Journal, 2008. **137**(2): p. 302-307.
4. Guo, Q.j., Y. Xu, and X. Yue, *Fluidization Characteristics in Micro-Fluidized Beds of Various Inner Diameters.* Chemical Engineering & Technology, 2009. **32**(12): p. 1992-1999.

507312 An Improved Bubble-Based Drag Model for Accurate Coarse-Grid Two-Fluid Modeling of Geldart a Powder Bubbling Fluidization

Abstract:

An improved bubble-based energy minimization multi-scale (EMMS) drag model was formulated in which the original bubble-based EMMS drag model at macroscale bed level was extended to microscale sub-grid level for accurate coarse-grid two-fluid modeling (TFM) of Geldart A powder bubbling fluidization. At both the macroscale and microscale levels, the physically heterogeneous structures were modeled by three pseudo phases, i.e., bubble phase, emulsion phase, and inter-phase, where conservation and stability condition equations were applied. Through such scale and phase decompositions, the improved bubble-based EMMS drag model was formulated through a two-step macro-to-micro scheme, where the obtained drag coefficient is related to phase volume fractions, velocities, and accelerations within each coarse grid. Modeling results show that TFM combined with the improved bubble-based EMMS drag model can accurately reproduce the heterogeneous structures for a laboratory-scale Geldart A powder bubbling fluidized bed. At relatively coarse grid, the predicted axial and radial distributions of solid concentration agree very well with experimental measurements. This indicates that our proposed improved bubble-based EMMS drag model incorporated TFM is promising for accurate and economical simulation of engineering-scale Geldart A powder bubbling fluidization.

507860 Numerical Simulation on Fine Particle Transport Behaviour in Electrostatic Precipitators

Abstract:

Electrostatic precipitators (ESPs) are a major air pollution control device to remove particles from treated gas. Extensive research has been done on several aspects such as electrostatics, fluid dynamics, charging mechanism, and particle dynamics. The overall particle-collection efficiency of a modern industrial wire-plate ESP exceeds 99%. However, the removal efficiency of submicron particles by conventional ESPs is as low as 70â€“80% owing to the low charge carried by fine particles and the increase in mobility with decreasing size. Submicron particles possess a high superficial area, thus favouring toxic heavy metals and other pollutants. Therefore, fine particles are collected with much lower efficiency and further research is required to improve this aspect.

In this work, a numerical study on the charging and transport behaviour of fine particles has been carried out based on wire-plate electrostatic precipitators (ESPs) with multiple wire electrodes. The effect of the applied wire voltage, inlet height, and wire distance on particle charging and transport behaviour, and the influence of the precipitator structure on particle trapping are analysed in detail. The 2-D numerical model involves electric field, space-charge density, gas flow, and particle trajectories. The fluid-flow velocity is higher than 1 m/s in this study and the EHD flow has a negligible. A Lagrangian approach for particle transport is utilized with the trajectories of individual particles tracked by solving their equation of motion. The electric field and gas flow are calculated using the finite-volume method. The flow simulation is done using the commercial software FLUENT with other parts coded as user-defined functions. The following main results can be drawn:

(1) It was observed that the Y-displacement of the particles exhibited a universal increase for all particle sizes, especially for larger particles, at low injection positions ($Y < 25$ mm). The larger the particle size, the higher the Y-displacement of particles because larger particle sizes improve both the particle charging level and electric field strength. No significant improvements were found for submicron particles.

(2) Increasing the spacing of the discharge electrode within a certain range and therefore increasing the electric field, it was observed that the particle-trapping effect improved and the outlet level exhibited a universal increase for all particle sizes, especially for large particles.

(3) The ESP models M1 and M2 were used to investigate their influence on the outlet height. Except for particles with a diameter of 2 μm, all other particle sizes showed an increased outlet height in M2 compared to that in M0. In addition, the different discharge-electrode arrangements, especially in M2, exhibited significant effects on the charging and transport behaviour of particles with a diameter of 1 μm.

(4) Particles with an inlet height of 5 and 10 mm were studied in M1 and M2 precipitators. The changed discharge-electrode arrangement improves particle trapping, especially in M2. This is because of the electron trajectories being closer to the discharge electrodes (M2), giving the particles more charge and leading to a greater electric field.

507879 Lattice Boltzmann Simulations of Porous Particulate Flows

Abstract:

The comprehensive understanding of the particle-fluid interaction is very important for fluid-particles system, and thus many studies have been conducted for the particle-fluid system[1-3] of solid impermeable particles. In the real industrial fields such as the catalyst process in chemical engineering[4], the particles usually possess the porous structure which will have effects on the fluid-particles interaction[5-6]. However, there has been little open literature discussing the particle-fluid system of porous permeable particles. Therefore, the objective of the present study is to investigate the effect of the porosity and permeability on the porous particulate flows which are formulated by the volume-averaged governing equations[7] in terms of intrinsic phase average velocity. The lattice Boltzmann equation model (LBE)[8] is used to solve the governing equations. Firstly, we validate the LBE model by the numerical results existing in the literature for the flow around and through one porous particle. Then, the porous particles settling against gravity in a quiescent fluid, i.e., the DKT process[9] for porous particles, is investigated. The effects of the porosity and Darcy number on the porous particulate flows are studied numerically in details. Also, we present the simulation results of the fluidization of 512 porous particles. Our results demonstrate the significance of the porous structure of particles for the particles dynamics behaviors.

Reference

[1] Ladd, A.J.C., 1994. Numerical simulations of particulate suspensions via a discretized Boltzmann equation Part I. Theoretical foundation. J. Fluid Mech., 271, 285-310.

[2] Hill, R.J., Koch, D.L., Ladd, A.J.C., 2001. The first effects of fluid inertia on flows in ordered and random arrays of spheres. J. Fluid Mech. 448, 213-241.

[3] Van der Hoef, M.A., Beetstra, R., Kuipers, J.A.M., 2005. Lattice Boltzmann simulations of low-Reynolds-number flow past mono- and bidisperse arrays of spheres: results for the permeability and drag force. J. Fluid Mech. 528, 233-254.

[4] Zhao, Y.F., Li, H., Ye, M., Liu, Z.M., 2013. 3D numerical simulation of a large scale MTO fluidized bed reactor. Ind. Eng. Chem. Res. 52, 11354-11364.

[5] Noymer, P.D., Glicksman, L.R., Devendran, A., 1998. Drag on a permeable cylinder in steady flow at moderate Reynolds numbers. Chem. Eng. Sci. 53, 2859-2869.

[6] Zhu, Q.Y., Chen, Y.Q., Yu, H.Z., 2014. Numerical simulation of the flow around and through a hygroscopic porous circular cylinder. Comput. Fluids 92, 188-198.

[7] Wang, L., Wang, L.-P., Guo, Z.L., Mi, J.C., 2015. Volume-averaged macroscopic equation for fluid flow in moving porous media. Int. J. Heat Mass Trans. 82, 357-368.

[8] Chen, S.Y., Doolen, G.D., 1998. Lattice Boltzmann method for fluid flows. Annu. Rev. Fluid Mech. 30, 329-364.

[9] Fortes, A.F., Joseph, D.D., Lundgren, T.S., 1987. Nonlinear mechanics of fluidization of beds of spherical particles. J. Fluid Mech. 177, 467-483.

508370 Modeling of a Novel Multi-Particle Collision Model for Gas-Solid Flows

Abstract:

Fluidized beds are widely applied in the chemical, petrochemical, metallurgical, environmental and energy industries due to their favourable mass and heat transfer characteristics. In past few decades computational fluid dynamics has proven to predict hydrodynamics of fluidized beds. DPM (Euler-Lagrange) [1] and TFM (Euler-Euler) [2] are two most common modelling approaches for gas-solid flow in fluidized beds. In TFM the solid phase itself is treated as a continuum phase interpenetrating with the gas phase. Since the concept of individual particles has disappeared, the effects of particle-particle interactions in the TFM has to be incorporated in a statistically averaged way, via the solid phase pressure and viscosity. TFM is preferred model to study large scale fluidized beds but model suffer from high numerical diffusion. And the KTGF theory in TFM assume elastic collisions and is limited only to spherical particles [2]. In DPM, the gas phase is considered as the continuous phase whereas the particle trajectories are obtained by integrating Newtonâ€™s equations of motion. Particle-particle and particle-wall interactions are explicitly taken into account using various physical models such as the soft sphere model. DPM proves to be accurate modelling of gas-particle flow. However, lagrangian models quickly become computationally too expensive because of the large number of particle interactions. Therefore coarse graining of discrete particles is an emerging model that can handle large number of particles effectively. The new coarse-grained model developments are of vital importance for the advancement of large-scale modelling of gas-particle flows that are flexible to extended for polydispersity, and non-spherical particles. In this study a novel lagrangian multi-particle collision (MPC) method is developed for predicting dense particulate-fluid flow.

The MPC solver is developed in OpenFoam [5], where the fluid phase is treated as a continuum and the solids are modelled by lagrangian particles. The MPC model grew out of the multiphase particle-in-cell (MP-PIC) model [3], that employs a fixed Eulerian grid, and particles are traced in a lagrangian fashion, including a parcel approach where each simulated particle represents a large collection of real particles. Each parcel follows Newtonâ€™s equations of motion, where inter-phase momentum transfer is accounted through mapping and interpolating quantities back and forth between parcel locations and the eulerian grid. Particle collisions in MP-PIC are not fully resolved but are derived from particles stresses that are theoretically calculated from Harris and Crighton [4] stress closure. In the MPC model, we implemented a functional form for the particle stress tensor where the coefficients are obtained computationally by means of DPM simulations. The idea is to extend the model from spherical particles to non-spherical particles, for which theoretical particles closures are currently not available in the literature. The viscous stress and random forces are typically ignored in MP-PIC model due to their insignificant effect [3]. Importantly, we incorporated viscous stress and random forces representing scattering of particles by collisions, which are obtained from DPM simulations by measuring the actual variance. Details of model development and validation is presented in this study.

Results from MPC model are compared with MPPICFoam and, with DPM simulations from DEM-LIGGGHT [6], and MFiX [7] code respectively. Pseuduo-2D fluidized beds of dimension 0.01x0.15x0.45 m is considered for comparison with experiments [8]. Comparison is made on the basic of pressure drop, void fraction, bubble dynamics and solids motion. We observed that MPC is in good agreement with DPM, and experiments, when compared to MPPICFoam. Unrealistic over-packing of solids is predicted in MPPIC due to limitation of Harris and Crighton [4] model and the velocity field updates, whereas in MPC these limitations overcomes and no over packing of particles is predicted. Solid circulations are in good agreement with DPM results, where solid moving above in the center and downward near to the wall. Time-average porosity plots revels the path of bubble motion and is in close agreement with results of DPM. In conclusion a novel multi-particle collision model is developed and validated to study gas-solid fluidized beds. The model is efficient to handle large number of particles. In future study, we plan to extend the model for dense flow of non-spherical particles.

Acknowledgments

The authors thank the European Research Council for its financial support under its consolidator grant scheme, contract no. 615906 (NonSphereFlow).

References

1. N.A. Patankar, D.D. Joseph, 2001. Modeling and numerical simulation of particulate flows by the Eulerian Lagrangian approach, International Journal of Multiphase Flow, Volume 27, Issue 10, October 2001, Pages 1659-1684.
2. Gidaspow, D., 1994. Multiphase flow and fluidization continuum and kinetic theory description, Academic Press, Boston.
3. Andrews, M.J., O'Rourke, P.J., 1996. The multiphase particle-in-cell (MP-PIC) method for dense particulate ows, Int. J. Multiphase Flow 22, 379-402.
4. Harris, S.E., Crighton, D.G., 1988. Solutions, solitary waves and voidage disturbances in gas-fluidized beds, J. Fluid Mech. 195, 437.
5. OpenFOAM, 2015. The Open Source CFD Toolbox, Ver. 3.0.0. ESI Group, Paris, France.
6. https://www.cfdem.com/liggghtsr-open-source-discrete-element-method-particle-simulation-code
7. https://mfix.netl.doe.gov/mfix/
8. M. Goldschmidt, 2001. PhD thesis, Hydrodynamic Modelling of Fluidised Bed Spray Granulation.

508490 Dpm Analysis of Large Fluidized Catalytic Cracking (FCC) Reactors

Abstract:

DPM Analysis of Large Fluidized Catalytic Cracking (FCC) Reactors

Abstract:

Computational Fluid Dynamic (CFD) modeling is a powerful tool for equipment design and development and is widely used at UOP, particularly for the reactor and regenerator design of FCC process unit. Through CFD modeling, a technical solution was developed for a large capacity FCC process unit purposing an elevated amount of propylene production.

A key challenge in FCC process design is how to optimize the size of the reactor vessel, which is a function of the number and size of reactor cyclones (catalyst separationdisengagers). UOP developed a new cyclone cluster configuration via CFD study of the reactor internals, including UOP's VSS™ riser termination device and reactor cyclones, which reduce the reactor size significantly compare to the conventional design, while maintaining effective catalyst / vapor separation and mechanical reliability. The objective of the current CFD study is to evaluate the catalyst separation efficiency and performance of the new riser termination device technology, which accommodate more cyclones in a smaller reactor vessel.

To better understand the flow pattern, flow distribution and the gas residence time inside the UOP VSS™ , the analysis was broken down into two sections. Part 1 focused on single gas phase only and in part 2 the particles behavior is studied using the Discrete Phase Model (DPM) in Fluent.

The single gas flow analysis results showed that the flow in the VSS chamber section has the same rotational pattern as in conventional FCC/VSS units, where the catalyst will separate from the gas products. Also the gas flow distribution at each chamber arm is uniform, which is critical to catalyst separation efficiency.

The DPM analysis confirmed an equal catalyst particle distribution through all the cyclones and a high efficiency separation in the riser termination device. Having this information from CFD allows for the new design to move forward.

509623 Catalytic Reactor Design Using Multiphase CFD

Abstract:

Catalytic Reactor Design Usng Multiphase Cfddimitri Gidaspow the 2011 Book , "Computational Gas-Solids Flows and Reacting Systems ", Edited By S. Pannala, M. Syamlal and Tom O'Brien ,Engineering Science Reference, Shows the Progress We Have Made in Using Multiphase CFD in Modeling Fluidized Bed Chemical Reactors and Other Multiphase Systems, Such As the Pyroplastic Flow Emitted from Volcanoes. in This Presentation the Design of Two Types of Catalytic Reactors Will be Described: A Fischer_ Tropsch Slurry Bubble Column Reactora Polymerization Reactor the CFD Design of a New Slurry Bubble Column Reactor Was Already Described By Gidaspow ,Et Al in Chemical Engineering Science 134,784-799 (2015).in This Presentation the Design Will be Extended to Obtain 100, 000 Barrels of Diesel per Day, Twice That of the Largest Commercial Reactors. the Production of Synthesis Gas Using Oxygen or Carbon Dioxide in Catalytic Fluidized Beds Needed for the Fischer-Tropsch Reactor Will Also be Discussed Briefly. for the Polymerization Reactor Our CFD Code Described in " Computational Techniques " By D. Gidaspow and V. Jiradilok , Nova Science, 2009 Was Modified for Polymerization Reactors. Such Reactors and Their CFD Modeling Are Described in a Chapter in Pannala's Et Al Book from Professor Rodney Fox's Group. They Point out That the Fine Particle (80 microns) Catalyst Radial Distribution, Is a Remaining Issue the Computations to be Presented Will Show That the Catalyst Will Have a High Concentration at the Reactor Walls for Various Operating Conditions. This Leads to a High Rate of Reaction at the Reactor Walls and May Result in Melting of the Polymers and Sheet Formation , Forcing Reactor Shut-Down. in Some of Our Simulations There Is Large Bubble and Slug Formation Similar to Those Mentioned in Fox's Chapter. Acknowledgement : I Thank Dr. Mayank Kashyap of SABIC for a Valuable Discussion and His Papers " Application of Multiphase Flow CFD in the Gas Phase Polymerization Process " Fluidization XV. May 24,2016 and a Related Paper Presented at Doe NETL, July 2016

509627 Simulation of Large Particle Turbulent Fluidization in Riser Reactors By Coarse Grain DEM-CFD

Abstract:

The complex two-phase flow in the riser is an important element of many industrial circulating fluidized bed reactors. In some cases, e.g. gas-solid polymerization, relatively large particles are transported upward in a turbulent regime, in which the high gas velocity results mostly from heat removal requirements. Strong solids acceleration, radial and axial velocity profiles for both the gas and solid, intense clustering and inhomogeneities make it impossible to reliably predict the riser head or local bed density profile. Detailed numerical simulations proved capable of showing some of the peculiarities of the flow, but limitations in the scale, lack of grid independence, need for special drag corrections and sub-grid models still hinder confident use of such tools in design and scale-up, particularly in industrial applications characterized by particle size and Reynolds numbers that are not well covered by the available scientific literature.

In the present contribution, coupled Discrete Element Method-Computational Fluid Dynamics (DEM-CFD) simulations are carried out using a coarse graining technique for the solid particles. According to this method, small number of particles are grouped into a computational particle (*parcel*) and represented by it as a spherical "lump," to reduce the computational cost of the simulations and allow lab-scale systems to be represented in full detail. The investigation focuses on the influence of the coars-graining degree, i.e. the number of real particles represented by a *parcel*, on the simulated flow hydrodynamics both in terms of its advantages (a quantification of the computational savings) and disadvantage, primarily the real capability of the method to resolve the behaviour observed and produce results in line with experimental findings. The gas and solid motion are studied, with the aim to characterize the clustering phenomena observed during the up-flow and its effect on the pressure drop. Assuming prescribed gas and solid inlet velocity fields, the potential tendency to show core-annulus flow, the intensity of solids clustering and the local bed density within the riser are examined at different number of particles per *parcel*.

509640 Numerical Study of Particles Shape Effects on Solid-Liquid Fluidizations

Abstract:

Numerical study of particles shape effects on solid-liquid fluidizations

Esmaeil Abbaszadeh Molaei, Zongyan Zhou*

Laboratory for Simulation and Modelling of Particulate Systems, Department of Chemical Engineering, Monash University, Clayton, VIC 3800, Australia

Solid-liquid fluidizations have been observed in many industries due to their high heat and mass transfer rate, high operation flexibilities, and reduced back mixing of fluid and solid phases. In most of these multiphase operations, particle properties such as size, density, and shape may experience significant changes during the process because of attrition, coalescence, comminution or chemical reactions. The difference in particle properties can result in significant segregation problem in liquid fluidized beds. To gain an understanding of solid-liquid fluidisation behaviour, most existing studies, either experimentally or numerically, have focused on solid flow behaviour with mono-sized particles or binary and ternary mixtures with different particle sizes or densities. In spite of substantial studies on the hydrodynamics of liquid fluidized beds, the effects of particles shape on the solid-liquid flow behaviour are still less reported. Except for two physical experiments, the previous studies considered particles as spheres, and few studies have been remarked on liquid fluidization of non-spherical particles. However, particles are generally irregular in most of the processes such as ore beneficiations using liquid-solid fluidized beds separator. Particle shape can cause different flow behaviour in liquid fluidizations rather than that observed for ideal spherical particles. Nonetheless, the answers to some fundamental questions such as how and why particle shape affects flow phenomena are not clear, and hence still poorly understood.

Therefore, in this work, a combination of computational fluid dynamics (CFD) and discrete element model (DEM) [CFD-DEM] is used to examine the effects of particle shape on the flow characteristics in liquid fluidizations. In the simulations, ellipsoidal particles are used as they can represent a wide range of particle shapes from oblate to prolate particles. Different aspect ratios are employed, and results are analysed mainly in terms of particle flow patterns, pressure drop, and bed expansion. In addition, force and particle orientation analysis is also conducted to examine the effect of particle shape and explain the occurrence of entrainment phenomenon.

The results show that ellipsoidal particles intend to enter to freeboard region and entrainment phenomenon occurs, and consequently, the interface between the bed and freeboard cannot be observed clearly. In addition, the pressure gradient variation with bed height is uniform for spheres; however, this uniformity is replaced with â€œdistorted parentheses-shapeâ€ profile for disc-shape and elongated particles. Finally, bed expansion analyses show that spherical particles have the lowest bed expansion. On the other hand, disc-shape particles intend to expand more significantly than spherical and elongated particles. This causes the entrainment phenomenon which has been found to be due to the effects of particles shape on particles terminal velocity.

* Corresponding author. Tel: 61 3 9905 0846; email: zongyan.zhou@monash.edu.

514318 CFD Investigation on Gas-Solids Flow and Heat Transfer in Two Fluidized Catalyst Cooler

Abstract:

Our previously experimental work showed that the internal circulation of solids in an annular catalyst cooler (ACC) had an ability to enhance the heat transfer between bed and wall, comparing with the base catalyst cooler (BCC). The current work simulated in detailed hydrodynamics of gas-solids flow and heat transfer to advance our understanding of the ACC by using the two-fluid model. The effective particle diameter d_pâŽ was determined to 170 Î¼m and the specularity coefficient Ï† was 0.3 in the 3D simulations. The catalyst cooler can be divided to four distinct regions including draft tube region, annulus region, bottom region and particle diffluence region according to the distributions of solids volume fraction and solids velocity. The effect of gas-solids flow on bed-to-wall heat transfer was compared in the different regions of the BCC and ACC. And the effects of operational gas velocity and bed structure were discussed. The results showed that the proper selection of the ratio of gas velocities and the position of the heat transfer tube were needed to form a stable internal solids circulation in the ACC. The ACC had a combined hydrodynamic feature of up-flow and down-flow catalyst coolers with bigger solids volume fraction and smaller particle resident time, which are beneficial for improving the heat transfer between solids and wall.

Hydrodynamics and mixing characteristics of a new-type particle mixer

Mengxi Liu, Chunxi Lu, Zhenliang Meng

State Key Laboratory of Heavy Oil Processing, China University of Petroleum, Beijing 102249, PR China

E-mail: mengxiliu@sina.com (M. Liu)

Abstract:

In order to further enhance the particle mixing in the internally circulating fluidized bed, especially along the radial direction, a new-type particle mixer with slots on the draft tube was proposed in the present work. The hydrodynamics of the mixer was investigated by using a multi-scale computational fluid dynamics with the EMMS drag. The prediction in terms of cross-sectionally averaged solid holdup and solid velocity were compared to experimental data for validation. The results show that in draft tube a low pressure region can be observed near the slots and in annulus the pressure in the vicinity of the slots seems larger than that in the region between the adjacent slots, which induces horizontal particle immigration and mixing. Nearly 25.4% to 29.7% of gas in the draft tube bypasses and flows through the slots into the annulus. The ratio of the mass flux of the tracer gas bypassing into the annulus to that of total tracer gas is approximately equal to 0. With increase of the superficial gas velocity in the draft tube from 0.3 m/s to 0.45 m/s, the total solid mass flux increases from 4.87 kg/s to 10.86 kg/s, while the mass flux through slots decreases from 2.69 kg/s to 0.52 kg/s.

1. Introduction

The characteristics of particle flow and mixing have a considerable influence on the mass and heat transfer, as well as the product quality. According to the previous study, the mixing degree of particles along the axial and radial direction is quite different [1-5]. Mostoufi and Chaouki [4] discovered that the diffusivities varied from 3.3×10^{-3} m^2/s to 5.6×10^{-2} m^2/s along the axial direction and 2.6×10^{-4} m^2/s to 1.5×10^{-3} m^2/s along the radial direction. Berruti et al. [6] also revealed that the particle mixing rate for radial direction was an order of magnitude smaller than for the axial direction. Therefore, the weak particle flow along the radial direction greatly limits the mixing ability of a fluidized bed. Internally circulating fluidized bed (ICFB) has been widely recognized as an efficient fluidized bed reactor, in which orderly and macroscopic particle circulation occurs. Compared with the conventional fluidized bed, ICFB offers multiple particle circulation between the draft tube and the annulus, which leads to relatively strong particle convection, diffusion and shear, and thereby improves the particle mixing performance along the radial direction [7]. In order to further enhance the particle mixing for some processes in which a uniform particle mixing is required, a new-type particle mixer with slots on the draft tube was proposed in the present work.

With increase of the computational power, the computational fluid dynamics (CFD) provides an additional tool to predict particle flow and mixing in the local region, which is difficult with the experimental methods. In particular, the gas-solid flow in a gas-solid fluidized bed can be accurately described by the energy minimization multi-scale (EMMS) model [8]. In this paper, the hydrodynamics of the mixer was investigated by using a multi-scale CFD based on the EMMS drag model [9, 10].

2. Experimental apparatus, Materials and Operating conditions

The structure of the particle mixer is shown in Fig.1. The mixer has a height of 5 m and an inner diameter of 286 mm. A draft tube with a height of 1 m and an inner diameter of 215 mm is coaxially mounted into the column. The four rectangular slots of 80×40 mm are symmetrically opened on the draft tube at a height of h=860 mm, in order to strengthen the particle radial flow. A gas distributor plate and a ring distributor are installed below the draft tube and the annulus respectively. The height of the gap between the gas distributor plate and the draft tube is 60 mm. For the sake of the convenience, the mixer is divided into six zones: the pre-mixing zone, feed influence zone, draft tube zone, gas-solid separation zone, annulus zone and the bottom zone. A perforated plate is mounted at the bottom of the pre-mixing zone, which is connected with a central downcomer, extending into the draft tube. Particles are first mixed in the pre-mixing zone and then the pre-mixed particles flow downward through the central downcomer and into the feed influence zone. In the feed influence zone, the pre-mixed particles are mixed with the particles moving upward in the draft tube, flow upward and circulate between the draft tube and the annulus. In the annulus, particles move downward to the bottom. Then a part of particles flow out of the mixer and the other part flow horizontally into the draft tube and recirculate.

The solid particles used in the experiment were FCC catalyst (ρ_p=1498 kg/m^3), with an average diameter of 76 μm. Air (ρ_g=1.225 kg/m^3) was introduced into the mixer to fluidize the particles. The superficial gas velocity ranged from 0.3 to 0.45 m/s in the draft tube (based on the cross-sectional area of the draft tube) and remained at 0.05m/s in the annulus (based on the cross-sectional area of the annulus). The mass flux of cold particles remained at 13.8 kg/m^{2}s. As we only investigated the influence of slots on the flow and mixing of particles in this paper, no hot particles were induced into the mixer.

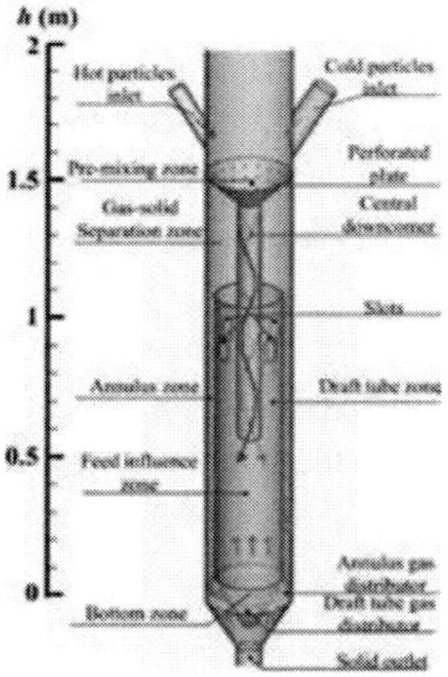

Fig. 1 Schematic diagram of the new-type particle mixer

3. CFD model and simulation method

The hydrodynamics of the mixer was performed using the Eulerian- Eulerian of Ansys Fluent

®13.0. The governing equations [11] include a set of mass and momentum equations for each phase, as shown in Table 1.

A 3-D mesh containing 580,000 cells was generated by using Gambit®2.4. The no-slip boundary was assigned for the gas phase and the partial-slip boundary condition developed by Johnson and Jackson [12] was used for the solid phase. In order to ensure a steady-state operation of the mixer, simulations were first performed for 30 s, and then the time-averaged data was analyzed for another period of 10 s. Other parameters settings in the model are shown in Table 2.

Table 1 Governing equations

Gas-phase Continuity equation	$\frac{\partial}{\partial t}(\varepsilon_g \rho_g) + \nabla\cdot(\varepsilon_g \rho_g \boldsymbol{u}_g) = 0$
Solid-phase Continuity equation	$\frac{\partial}{\partial t}(\varepsilon_s \rho_s) + \nabla\cdot(\varepsilon_s \rho_s \boldsymbol{u}_s) = 0$
Gas-phase momentum equation	$\frac{\partial}{\partial t}(\varepsilon_g \rho_g \boldsymbol{u}_g) + \nabla\cdot(\varepsilon_g \rho_g \boldsymbol{u}_g \boldsymbol{u}_g) = -\varepsilon_g \nabla p + \nabla\cdot\boldsymbol{\tau}_g + \varepsilon_g \rho_g \boldsymbol{g} - \beta(\boldsymbol{u}_g - \boldsymbol{u}_s)$
Solid-phase momentum equation	$\frac{\partial}{\partial t}(\varepsilon_s \rho_s \boldsymbol{u}_s) + \nabla\cdot(\varepsilon_s \rho_s \boldsymbol{u}_s \boldsymbol{u}_s) = -\varepsilon_s \nabla p - \nabla p_s + \nabla\cdot\boldsymbol{\tau}_s + \varepsilon_s \rho_s \boldsymbol{g} + \beta(\boldsymbol{u}_g - \boldsymbol{u}_s)$
Granular temperature equation	$\frac{3}{2}[\frac{\partial}{\partial t}(\varepsilon_s \rho_s \Theta_s) + \nabla\cdot(\varepsilon_s \rho_s \Theta_s \boldsymbol{u}_s)] = (-p_s \boldsymbol{I} + \boldsymbol{\tau}_s):\nabla \boldsymbol{u}_s + \nabla\cdot(\kappa_s \nabla \Theta_s) - \gamma_s - 3\beta\Theta_s$
Gas-phase stress	$\boldsymbol{\tau}_g = \varepsilon_g \mu_g [\nabla \boldsymbol{u}_g + (\nabla \boldsymbol{u}_g)^T] - \frac{2}{3}\varepsilon_g \mu_g (\nabla\cdot\boldsymbol{u}_g)\boldsymbol{I}$
Solid-phase stress	$\boldsymbol{\tau}_s = \varepsilon_s \mu_s [\nabla \boldsymbol{u}_s + (\nabla \boldsymbol{u}_s)^T] + \varepsilon_s(\lambda_s - \frac{2}{3}\mu_s)(\nabla\cdot\boldsymbol{u}_k)\boldsymbol{I}$
Solid-phase pressure	$p_s = \varepsilon_s \rho_s \Theta_s [1 + 2(1+e_s)\varepsilon_s g_0]$
Solid-phase shear viscosity	$\mu_s = \mu_{s,col} + \mu_{s,kin} + \mu_{s,fr}$
Where	$\mu_{s,col} = \frac{4}{5}\varepsilon_s \rho_s d_p g_0 (1+e_s)\sqrt{\Theta_s/\pi}$, $\mu_{s,kin} = \frac{10\rho_s d_p \sqrt{\Theta_s \pi}}{96\varepsilon_s(1+e_s)g_0}[1 + \frac{4}{5}g_0\varepsilon_s(1+e_s)]^2$, $\mu_{s,fr} = \frac{p_s \sin\phi}{2\sqrt{I_{2D}}}$
Radial distribution function	$g_0 = [1 - (\varepsilon_s / \varepsilon_{s,max})^{1/3}]^{-1}$
Diffusion coefficient	$\kappa_s = \frac{150\rho_s d_p \sqrt{\Theta_s/\pi}}{384(1+e_s)g_0}[1 + \frac{6}{5}g_0\varepsilon_s(1+e_s)]^2 + 2\rho_s\varepsilon_s^2 d_p (1+e_s) g_0 \sqrt{\Theta_s/\pi}$
Collisional energy dissipation	$\gamma_s = \frac{12(1-e_s^2)\varepsilon_s^2 \rho_s g_0 \Theta_s^{3/2}}{d_p\sqrt{\pi}}$
Heterogeneity index	$H_d = \frac{\beta_e}{\beta}$, $\beta_e = \beta H_d$
Drag based homogeneous structure	$\beta = \begin{cases} \frac{3}{4}C_d \frac{\varepsilon_s \varepsilon_g \rho_g \lvert \boldsymbol{u}_g - \boldsymbol{u}_s \rvert}{d_p}\varepsilon_g^{-2.65}, \varepsilon_g > 0.8 \\ 150\frac{\varepsilon_s^2}{\varepsilon_g}\frac{\mu_g}{d_p^2} + 1.75\varepsilon_s \frac{\rho_p}{d_p}\lvert \boldsymbol{u}_g - \boldsymbol{u}_s \rvert, \varepsilon_g \le 0.8 \end{cases}$
Where	$C_d = \begin{cases} 0.44 & Re_s \ge 1000 \\ \frac{24}{Re_s}(1 + 0.15 Re_s^{0.687}), & Re_s < 1000 \end{cases}$, $Re_s = \frac{\varepsilon_g \rho_g d_p \lvert \boldsymbol{u}_g - \boldsymbol{u}_s \rvert}{\mu_g}$

Table 2 Simulation Parameters

Gas-solid model	Eulerian-Eulerian
Solids Pressure	Lun et al.
Particle-wall specularity coefficient	0.5
Gas-wall friction coefficient	No-slip
Pressure-Velocity Coupling	Simple
Gradient discretization	Least squares cell based
Momentum discretization	Second-order upwind
Volume fraction discretization	Quick
Max. number of iterations per time step	40
Pressure outlet (atm)	1
Inventory (kg)	56
Incipient solid holdup	0.6

4. Results and discussion

4.1 Model validation

Fig. 2 gives the distribution of cross-sectionally averaged solid holdup and particle velocity in the mixer. As shown in Fig. 2(a), the solid holdup is larger in the bottom zone (h=0.04 m) and due to the accelerated upward motion of particles, the solid holdup decreases with increase of the axial height (when $h \leq 0.16$ m). The cross-sectionally averaged solid holdup is nearly constant in the draft tube and gas-solid separation zone of $0.4 \leq h \leq 1.25$ m. The cross-sectionally averaged particle velocity is shown in Fig. 2(b). The particle velocity is smaller in the bottom zone and keeps almost invariable in the draft tube and the gas-solid separation zone. Compared with the experimental data, the particle velocity seems slightly higher, while the solid holdup in the simulation agrees reasonably with the experimental results.

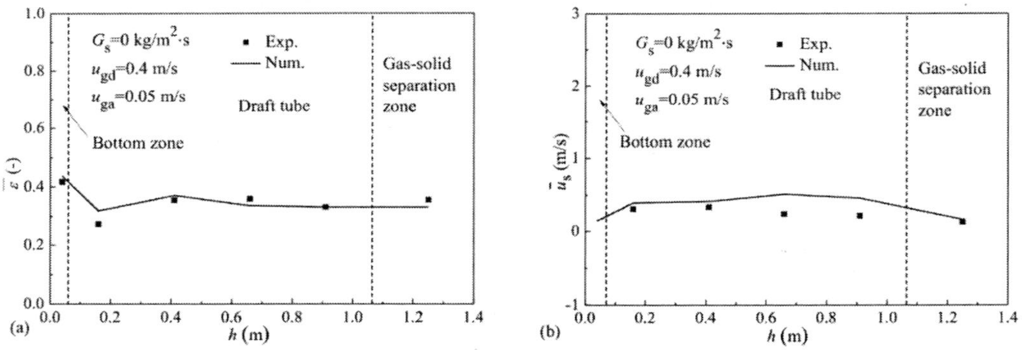

Fig. 2 Distribution of cross-sectionally averaged solid holdup and particle velocity as a function of axial heights

4.2 Pressure distribution near the slots

Fig. 3 shows the time-averaged contours of static pressure on the cross-section of h=0.9 m, the cross-section on the middle of the slots. It is seen that in the draft tube a low pressure region can be observed near the slots, smaller than the pressure in the region between the adjacent slots. This pressure drop forced particles inside the draft tube to move towards the slots and pass through them, which significantly facilitates the mixing of the particles along the radial direction. In the annulus the pressure in the vicinity of the slots seems larger than that in the region between the adjacent slots, inducing horizontal particle immigration and mixing.

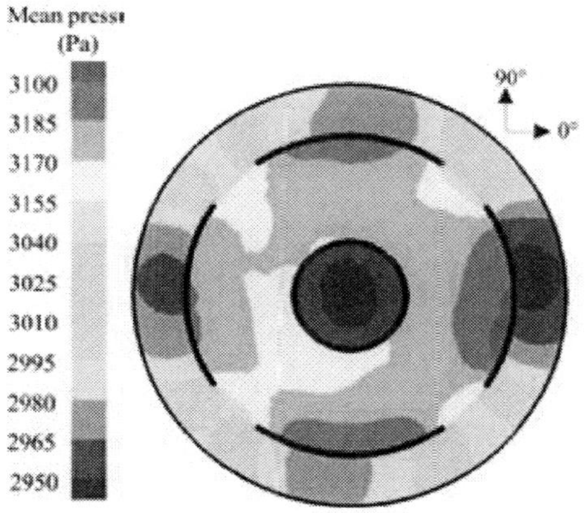

Fig. 3 Time-averaged contours of static pressure at the cross-section of the slots (u_{gd} =0.4 m/s, G_s=13.8 kg/m^{2}s, h=0.9 m)

4.3 Gas bypassing

In the present work, the gas bypassing behavior was investigated by CFD simulation, and a tracer gas having identical physical properties with air was defined. After the mixer running stably, the tracer gas was introduced into the mixer from the gas distributor plate to replace the original gas. A cross section (h=0.06 m) on the exit of the annulus was selected and the mass flux of the tracer gas through the cross section was registered. The results are shown in Fig. 4. f_B refers to the ratio of the mass flux of the tracer gas through the cross section (h=0.06 m) to that of total tracer gas. It is seen that f_B varies between -0.05 and 0.05 with a frequency of 10 Hz, indicating an unstable gas bypassing from the draft tube to annulus. This suggests that the gas bypassing into the annulus was finally brought out of the annulus by the circulating particles. The average f_B is approximately equal to 0, much smaller than that in a conventional ICFB (usually within 15%) [13], because massive particle circulation in the mixer makes it difficult to bypass from the draft tube to annulus.

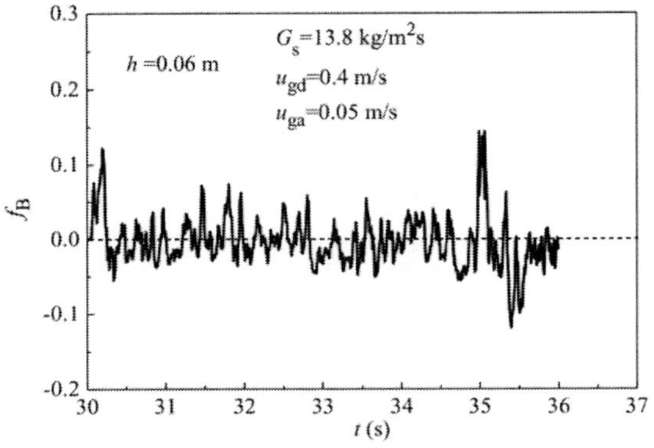

Fig. 4 Gas bypassing from draft tube to annulus in the bottom zone

As discussed previously, in the mixer there exists a pressure gradient from the draft tube to the vicinity of the slots, and then the annulus. Such a pressure difference leads to horizontal flows of particles and gas from the draft tube to the slots and then to the annulus. In the mixer the upside and downside of the slots is located at h=0.94 m and 0.86 m, respectively. Mass flux of the gas bypassing through the slots was obtained by monitoring the gas flux through the lower cross section (h=0.7 m) and the upper cross section (h=1 m) in the draft tube. Fig. 5 shows the gas mass flux under different operating conditions. The total gas mass flux in the draft tube increases with increasing gas velocity in the draft tube. With increase of the gas velocity from 0.3 to 0.45 m/s, the mass flux of the gas bypassing through the slots increases from 0.157 to 0.221 kg/m^{2}s. But the mass flux of bypassing gas does not increase linearly with the gas velocity. It is also seen that the mass ratio of the gas through the slots to that in the draft tube drops from 29.7% to 25.4%.

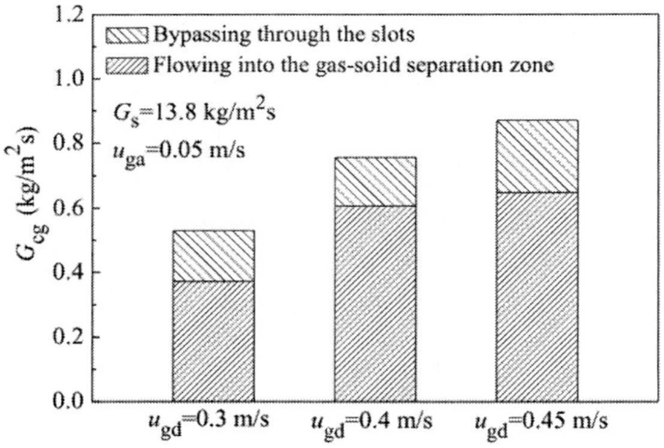

Fig. 5 Gas mass flux through the slots and the gas-solid separation zone as a function of superficial gas velocity in the draft tube

4.4 Particle mass flux

Fig. 6 shows the particle mass flux in the mixer. It is seen that the mass flux increases with increase of the superficial gas velocity in the draft tube. The mass flux through the slots accounts for about 50% of the total mass flux when u_{gd}=0.3 m/s. Because the axial particle velocity rose with increase of the superficial gas velocity from 0.3 m/s to 0.45 m/s in the draft tube, making it difficult to change the direction of the particles. Meanwhile, the radial particle velocity above the draft tube increases. Consequently, with the increase of the superficial gas velocity in the draft tube , the mass flux through the slots considerably decreased, from about 55.3% of the total mass flux (about 2.69 kg/s) at u_{gd}=0.3 m/s to only 4.8% (about 0.52 kg/s) at u_{gd}=0.45 m/s. Meanwhile, the total particle mass flux in the mixer increased from 4.87 to 10.86 kg/s as u_{gd} increased from 0.3 to 0.45 m/s.

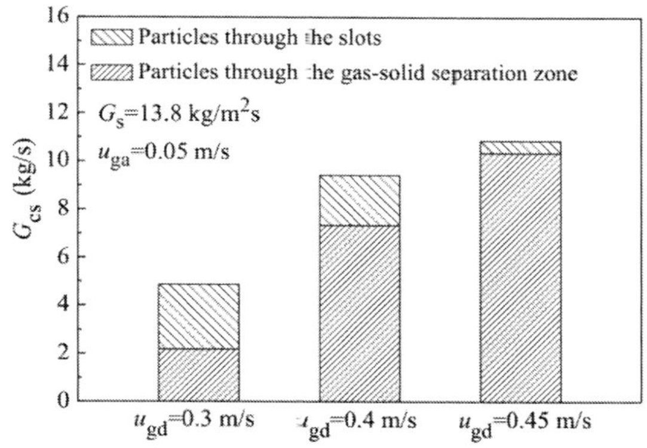

Fig. 6 Particle mass flux through the slots and the gas-solid separation zone as a function of superficial gas velocity in the draft tube

5. Conclusions

The characteristics of particle flow and mixing have a considerable influence on the mass and heat transfer, as well as the product quality. In order to further enhance the particle mixing in the ICFB, especially along the radial direction, a new-type particle mixer with slots on the draft tube was proposed in the present work. The hydrodynamics of the mixer was investigated by using a multi-scale computational fluid dynamics with the EMMS drag. The prediction in terms of cross-sectionally averaged solid holdup and solid velocity were compared to experimental data for validation. The results show that in draft tube a low pressure region can be observed near the slots and in annulus the pressure in the vicinity of the slots seems larger than that in the region between the adjacent slots, which induces horizontal particle immigration and mixing. Nearly 25.4% to 29.7% of gas in the draft tube bypasses and flows through the slots into the annulus. The ratio of the mass flux of the tracer gas bypassing into the annulus to that of total tracer gas is approximately equal to 0. With increasing the superficial gas velocity in the draft tube from 0.3 m/s to 0.45 m/s, the total solid mass flux increases from 4.87 kg/s to 10.86 kg/s, while the mass flux through slots decreases from 2.69 kg/s to 0.52 kg/s.

Notation

G_s	solid mass flux (kg/m^2 s)	G_{cg}	gas mass flux in the mixer (kg/m^2 s)
G_{cs}	solid mass flux in the mixer (kg/s)	h	height (m)

p	pressure (Pa)	R	radius of the reactor (mm)
r	radial position of measurement points (mm)	t	time (s)
u_{ga}	superficial gas velocity in the annulus (m/s)	u_{gd}	superficial gas velocity in the draft tube (m/s)
$\bar{u}$	cross-sectionally averaged axial particle velocity (m/s)		

Greek

ε_g	voidage	ε_s	solid holdup
$\bar{\varepsilon}_s$	cross-sectionally averaged solid holdup	ρ	density (kg/m^3)

References

[1] F. Niklasson, H. Thunman, F. Johnsson, B. Leckner, Estimation of solids mixing in a fluidized-bed combustor, Ind Eng Chem Res, 41 (2002) 4663-4673.

[2] L.H. Shen, M.Y. Zhang, Effect of particle size on solids mixing in bubbling fluidized beds, Powder Technol, 97 (1998) 170-177.

[3] L.H. Shen, M.Y. Zhang, Y.Q. Xu, Solids Mixing in Fluidized-Beds, Powder Technol, 84 (1995) 207-212.

[4] N. Mostoufi, J. Chaouki, Local solid mixing in gas-solid fluidized beds, Powder Technol, 114 (2001) 23-31.

[5] D.Y. Liu, X.P. Chen, Lateral solids dispersion coefficient in large-scale fluidized beds, Combust Flame, 157 (2010) 2116-2124.

[6] F. Berruti, D.S. Scott, E. Rhodes, Measuring and Modeling Lateral Solid Mixing in a 3-Dimensional Batch Gas Solid Fluidized-Bed Reactor, Can J Chem Eng, 64 (1986) 48-56.

[7] M.M. Fang, K. Luo, S.L. Yang, K. Zhang, J.R. Fan, Computational Fluid Dynamics-Discrete Element Method Investigation of Solid Mixing Characteristics in an Internally Circulating Fluidized Bed, Ind Eng Chem Res, 52 (2013) 7556-7568.

[8] J. Li, M. Kwauk, Particle-fluid Two-phase Flow: the Energy-minimization Multi-scale Method, Metallurgical Industry Press, Beijing, 1994.

[9] Z.S. Shi, W. Wang, J.H. Li, A bubble-based EMMS model for gas-solid bubbling fluidization, Chem Eng Sci, 66 (2011) 5541-5555.

[10] K. Hong, Z.S. Shi, W. Wang, J.H. Li, A structure-dependent multi-fluid model (SFM) for heterogeneous gas-solid flow, Chem Eng Sci, 99 (2013) 191-202.

[11] D. Gidaspow, Multiphase flow and fluidization: Continuum and Kinetic Theory Descriptions, Academic Press, Boston, 1994.

[12] P.C. Johnson, R. Jackson, Frictional–collisional constitutive relations for granular materials, with application to plane shearing, J. Fluid Mech. 176 (1987) 67–93.

[13] B.H. Song, Y.T. Kim, S.D. Kim, Circulation of solids and gas bypassing in an internally circulating fluidized bed with a draft tube, Chem Eng J, 68 (1997) 115-122.

514498 Power Spectral Density Analysis of Pressure Signal in 18 m Circulation Fluidized Bed Riser

Abstract:

Studies on the gas-solids flow characteristics using Power Spectral Density based on the pressure data from different axial locations in a riser of diameter 80 mm and height 18 m. Results show that the pressure signal fluctuation is mainly affected by gas-solids interaction, which included gas-solids friction, particle-particle collision and particle-particle aggregation. With the increasing solids circulating rate, frequency of particle interactions including particle-particle collision and particle-particle arrogation become strong and the pressure fluctuation also increased. It can be seen that the pressure signal fluctuation intensity decreased gradually with the increasing axial position of the riser. The low frequency but high power phenomenon nearly disappears in the fully developed zone near the top of the riser. Based on the results, it can be concluded that the gas-solids interaction is stronger at the bottom of riser. On the other hand, the gas-solid interaction is more homogeneous, and the pressure fluctuation is relatively weak in the fully developed zone.

514514 Characteristics of Pressure Fluctuations in Particle-Transport Inclined Pipe of a Circulating Fluidized Bed

Abstract:

In a large circulating fluidized bed device, the characteristics of pressure fluctuation in a particle-transport inclined pipe under different solid circulating flux were investigated based on measuring the dynamic pressure of gas-solid two-phase flow by using FCC equilibrium catalyst as the particles. The experimental results showed that with increase of particle mass flux, the flow patterns of FCC particles above and below the butterfly valve in the inclined pipe both successively changed, and the fluctuation amplitude of dynamic pressure increased. The standard deviation of the dynamic pressure below the butterfly valve possessed a linear relation with the particle mass flux, which could be used to characterize the real-time dynamic pressure in the inclined pipe. By wavelet analysis, it is indicated that there exists a main frequency caused by inclined pipe flow structures in the pressure fluctuations above and below the butterfly valve, besides, there is a secondary frequency in the pressure fluctuation below the butterfly valve caused by gas turbulence and gas-solid interactions. With the increase of particle mass flux, the wavelet energy ratio of the secondary frequency gradually decreased. The fluctuation pressure caused by discharge in the inclined pipe is a low frequency pressure and the excitation source of inclined pipe vibration.

508323 Dynamic Modeling of Attrition and Reactions in Circulating Fluidized Bed Reactors

Abstract:

Fluidized bed reactors are widely applied in chemical processes due to the excellent gas-solid contact. A problem that occurs in fluidized beds is the mechanical stress that particles are exposed to during the fluidization, which causes particle attrition. Additionally, fluidized beds are often operated in batch mode, which will cause ever-changing operating conditions with time. This makes fluidized bed modeling especially challenging, because the distributed particle properties, like size, wetness or reaction state, have to be tracked over the whole course of the simulation.

At the Institute of Solids Process Engineering at Hamburg University of Technology, novel process simulation software named DYSSOL is developed, particularly tailored for solids processes. The software is based on the flowsheeting concept. This means that chemical unit operations, like for example chemical reactors or separators, are put into process modules, which are interconnected into a process flowsheet and then mathematically solved. To make dynamic simulations available, new dynamic models had to be proposed for the involved equipment. The models themselves are based on empirical and semi-empirical correlations for hydrodynamics and kinetic particle reaction models are integrated.

The dynamic flowsheeting software was then used to simulate a system of interconnected fluidized bed reactors, which are used for Chemical Looping Combustion of solid fuels. This process is especially interesting, because heterogeneous reactions take place. Further, the process is a reactor-regenerator system causing the reactive solid to be oxidized and reduced and, hence changing its conversion state. Attrition plays a major role for the cost effective operation of the process.

In a small lab scale bubbling bed reactor, the jet and bubbling attrition rate constants for the solid oxygen carrier were experimentally determined under ambient conditions. In a small stand-alone cyclone the cyclone attrition was investigated. Furthermore, the reaction kinetics of the solid material with fuel gases was experimentally determined. All this information was included into the flowsheet simulation of a pilot scale circulating fluidized bed reactor, which is actually operated in our institute to deliver validation data.

The dynamic flowsheet simulation of the pilot plant, which was enhanced with kinetic and attrition information of smaller scale facilities, was conducted over 1000 seconds. A fuel load change was simulated and the chemical conversion was tracked. Over the whole simulation, the attrition and small particles leaving the reactor system was also covered. When comparing the simulation with experiments it could be shown that the gas conversion was sufficiently calculated and the conversion state of the oxygen carrier tracked correctly. The attrition modeling, however, with its parameters measured had to be adapted, so that it fitted the outlet particle flows measured under reaction conditions.

The financial support of DFG (Deutsche Forschungsgemeinschaft) within the priority program SPP 1679: â€œDynamic simulation of interconnected solids processesâ€ is gratefully acknowledged; grant number HA 6935/2-1.

514599 Analysis of FCC Cyclone Fault Diagnosis Technology Based on Particles Information

Abstract:

Cyclones are key equipment in fluid catalytic cracking unit (FCCU), which are used for separating the catalyst particles from oil gas or flue gas. Hence, it is very necessary to carry out condition monitoring and fault diagnosis for the running cyclone. The properties of particles vary at different positions in the particle circulation loop, such as size distribution, morphology and element composition, etc. In consequence, the particles contain much information about operation. Once the cyclone operates abnormally, the information of run-off catalyst particles appear abnormal too, and there exists a corresponding relationship between them. So, taking the catalyst particles as the carrier of fault information, this paper put forward cyclone fault diagnosis technology by extracting catalyst particle information, such as particle packing density, particle size distribution, particle micrographs etc, to build the relationship model between the cyclone fault and catalyst particle information.

514600 The Multi-Hole Throttling Effect and the Erosion Characteristics of the High Pressure Natural Gas

Abstract:

Based on Euler-Lagrange multi-phase flow numerical simulation method, the single hole and multi-hole throttling process of the high pressure natural gas was studied by using the RNG k-$\hat{I}\mu$ turbulent model, the Sutherland Viscosity Law compressible fluid viscosity model, and the advanced erosion model numerical simulation. The calculated mass flow variation law in the throttle choke is comparable with the theoretical calculations, which verified the adopted models and calculating methods. And then the throttling effect and the erosion characteristic of the high pressure natural gas was studied to check its specific law with different throttle holes by using the advanced erosion model. The results showed that, high pressure natural gas can reduce pressure and expansion after flowing through the single hole and multi-hole throttle choke, in which lead a high-speed jet region came up near its outlet. But the maximum jet velocity of multi-hole throttle choke was significantly lower than that of the single hole. At the same time the multi-hole uniform jet may reduce the vortex entrainment and the wall erosion damage, in which can be easier to mix the cold core and achieve stable pressure recovery at the role of the multi-jet mixing effect. Therefore, the multi-hole throttling can effectively restrain the erosion damage and the gas hydrate formation probability.

514601 Numerical Simulation of Flow Field In a Gas Pipe Distributor of the FCCU Regenerator

Abstract:

The flow field in the pipe distributor used in regenerator of FCCU (Fluid Catalytic Cracking Unit), was analyzed based on the numerical simulation. Firstly, the distributing gas performance of the pipe distributor was investigated. The result shows that the pressure increases along the branch pipe of the pipe distributor and the axial flow rate decreases in the branch pipe, which leads to the nozzle flow rate increasing gradually. Furthermore, an obvious bias flow was found in the inlet of branch pipe, which brings some nozzle flow rate less than the designed flow rate and causes an uneven gas distribution in the fluidized bed as well. Meanwhile, the momentum equation with variable mass flow was applied to analyze the flow in the pipe distributor. The flow type in the branch pipe belongs to the "momentum exchange controlling model". The pressure in the branch pipe always increases from the inlet to the end, and the pressure distribution is not consequentially uniform. The uneven gas distribution of the pipe distributor results in fluidized gas non-uniformity in the bed and erosion in the branch pipe and nozzle. Finally, according to the pressure characteristics in fluidized bed system, relationship between the pressure drop of branch pipe and pressure drop of nozzles, as well as of the critical nozzle pressure drop and the designed nozzle pressure drop, were discussed. The gas distributor performance could be optimized by means of the structural improvement of branch pipe.

514602 Full-Loop Simulation of Gas-solids Flow in Circulating Fluidized Bed of FCC System

Abstract:

Circulating fluidized beds are widely used in petrochemical, coal, metallurgy, and other industrial fields. A typical circulating fluidized bed (CFB) includes the riser, gas-solid separator, downer and solids flow control device. To obtain detailed information and get better understand of gas-solids flow in CFB, computational fluid dynamics (CFD) had been applied extensively in many systems such as circulating fluidized bed combustion (CFBC) systems and fluid catalytic cracking (FCC) systems. In most studies, the riser, the core part in CFB system, was isolated as a single simulation object by applying proper boundary conditions. However, there are some limitations of riser-only simulation, without consideration of dynamics in solids circulation rate and effects of other part of CFB system. Therefore, the full-loop simulation is a more promising way to describe complex gas-solids interaction in CFB by treating the whole CFB system as a simulation object. Nowadays, most of current full-loop simulations were focused on CFBC systems, there was little work about FCC system, with different type of particles and recycle mechanism from CFBC system.

In this work, both full-loop and riser-only simulations were carried out by TFM approach to investigate a pilot-scale gas-solid circulating fluidized bed of FCC system which has been set up in our group. The simulation condition was chosen according to experiments with different superficial gas velocity, valve opening and fluidizing air flow rate. Detailed comparison between full-loop and riser-only numerical simulations has been conducted with respect to the hydrodynamics inside the riser, which showed full-loop simulation had advantages on predicting dynamics of solids circulation rate and the gas-solids flow behavior. Distribution of pressure, solids holdup and velocity in system was investigated which was helpful to understand the hydrodynamics in CFB. The effects of operating condition were then studied, which revealed the system stability mechanism.

506758 Immobilization of Sulfur-Oxidizing Bacterium, *Thioalkalivibrio Sp.* D301 on Magnetic Nanoparticles and Biodesulfurization

Abstract:

There is an urgent need to develop desulfurization methods for halo-alkaline wastewater. *Thialkalivibrio sp.* D301, an obligatory chemolithoautotrophic, haloalkaliphilic, sulfur-oxidizing bacterium, was used in the removal of sulfide in wastewater. We developed a process to immobilize D301 cells by using superparamagnetic Fe_3O_4 nanoparticles (NPs) under halo-alkaline conditions. Fe_3O_4 NPs were synthesized by coprecipitation of ferric and ferrous salts in the presence of N_2 gas. TEM image shows the spherical shape of Fe_3O_4 NPs with an average diameter of 10 to 15 nm. The particles had superparamagnetic properties. The saturation magnetization (Î´$_s$) of the Fe_3O_4 NPs was 73.985 emu g$^{âˆ'1}$ with zero remanence and coercivity. Analysis of zeta potentials and hydraulic diameters showed that the Fe_3O_4 NPs and bacterial cells interact mainly through the nano-size effect. The NPs-coated cells have the same sulfur oxidization activity as that of free cells. Experiment results showed that the immobilized cells still retained >90% of their sulfur oxidization activity after six times of reuse. The cells coated with Fe_3O_4 NPs did not interact through electrostatic attraction. The mechanism of NPs adsorption mainly involved the nano-size effect. Magnetic immobilization is a convenient technique for HA-SOB immobilization under halo-alkaline conditions and is a promising technology for large-scale industrial applications.

507944 Simulation of Bulk Solids and Granular Systems By Using Combined Discrete Element Models

Abstract:

Granular materials can be found in many industrial processes from chemical to energy, agriculture, mining, pharmacy and many other fields. They are almost always opaque, making the acquisition of experimental data below the surface difficult. DEM (Discrete Element Method) simulations are an effective way to study the physics of granular materials because they can simulate the micro-dynamic behavior of granular material on an individual particle scale and the motion of every particle can be traced. Since DEM was first considered, many modified discrete element models have been developed. However, most existing DEM simulation are just for spherical particles or a specific type of non-spherical particles. In this work, a combination of several types of discrete element models including spherical models, super-ellipsoids, multi-sphere models (with no interaction and deformation between sub-spheres), and bonded-spheres models (with interaction and deformation between sub-spheres) were used to simulate the flow of granular materials with many different shapes. Several simulation cases including the flow of particles in tumblers, blenders, and other devices were performed to validate the modeling approaches. The simulation results show that it is practical and efficient to combine different discrete element models together for simulating both spherical and non-spherical particles in bulk solids and granular systems.

515703 Material Optimization of Perovskite Films using High Throughput Synthesis and Multi-Dimensional Analysis

Abstract:

Over the past years, Perovskites received special attention in the field of renewable energy as they are one of the prime candidates for energy storage and conversion materials. Perovskites act as the active light harvesting layer of a solar cell and provide excellent performance.[1] A common synthesis strategy is through the liquid phase under inert atmosphere.[2,3] However, this comes with the challenge of a huge parameter space that in principle would have to be first synthesized, then characterized and finally optimized based on the therefrom deduced knowledge.

To overcome these limitations, we follow a fully automated approach, which allows automated high throughput (HT) synthesis and HT characterization of perovskites with high accuracy. As an efficient route to explore the parameter space, we use design of experiments (DoE) to find optimized settings for process parameters such as synthesis and annealing time and temperature, but also material aspects like the molar ratio of the reactants.

For all experiments, a Chemspeed© technologies (CST) swing XL platform with a liquid handling tool and a heatable shaking rack was used. The robotic platform is enclosed within an Mbraun© glovebox with inert atmosphere (Nitrogen). After precursor preparation, deposition was realized by a liquid handling tool on glass substrates inserted in a customized aluminum 24 well plate, which is then transferred onto a hot plate for film formation.

In a first step, after annealing at different temperatures, a characterization cascade was developed and applied that contains the subsequent elimination of samples that exhibit undesired properties and side products. Fast absorption analysis, photoluminescence (PL) measurements and finally X-ray diffraction (XRD) were performed to identify optimum chemistry and process parameters. Although this is a very reliable procedure, it is still comparatively demanding in terms of the time effort. Therefore a more efficient, purely optical approach by which the experimental parameter space is only explored in terms of rapid emission analysis was developed in a second step. Noteworthy, both strategies led to the same optimum conditions, however, at significantly reduced time effort.

In conclusion, our work shows how by the combination of automation, DoE and efficient characterization, optimum conditions for new materials like perovskites are rapidly identified. This opens the door towards faster material development which is important whenever new perovskite structures, e.g. lead free materials for industrial applications, are strived for.

Literature

1. Green, M.A.; Ho-Bailie, A.; Snaith, H.J., "The emergence of perovskite solar cells", Nature Photonics 2014, 8, (7), 506-514.
2. Baikie, T.; Fang, Y.; Kadro, J.M.; Schreyer, M.; Wei, F.; Mhaisalkar, S.G.; Graetzel, M.; White, T.J., "Synthesis and crystal chemistry of the hybrid perovskite (CH 3NH3)PbI3 for solid-state sensitised solar cell applications", J. Mater. Chem. A 2013,1 , 5628-5641.
3. Yan, K.; Long, M.; Zhang, T.; Wei, Z.; Chen, H.; Yang, S.; Xu, J., "Hybrid Halide Perovskite Solar Cell Precursors: Colloidal Chemistry and Coordination Engineering behind Device Processing for High Efficiency", J. Am. Chem. Soc. 2015, 137, (13), 4460–4468.

509509 Solid Liquid Separation Via Particle Flow Instability

Abstract:

A new separation strategy using novel particle-flow-instability physics is successfully developed for a *difficult-to-separate suspension* in which there is some combination of a small density difference between solid and liquid, high viscosity, and small-sized particles. The method we propose here requires no dilution for medium, high or extremely high viscosity slurries and can produce effective solidâ€"liquid separation with low capital and operational inputs. The results have relevance for many biotechnological and/or chemical processes.

509543 Milling and Grindability Assessment of Pharmaceutical Materials

Abstract:

Milling is one of the most important unit operations in the pharmaceutical industries. Most of the Active Pharmaceutical Ingredients (APIs) and excipients are produced by crystallisation, and grown to a sufficiently large particle size suitable for isolation and drying. Therefore most of the materials have to be milled to desirable particle sizes. Milling is influenced by a combined effect of milling dynamics (input power, mill geometry, etc) and material properties. Considerable research has been carried out on analysing the particle behaviour in various mills. Nevertheless, the prediction of the milling rate and milled product properties (e.g. product size distribution) is still very challenging.

In this work, the milling behaviour of several pharmaceutical materials in the PicoPlex mill (pin mill unit of the Picoline, Hosokawa Alpine) has been studied. A combination of experiments and Distinct Element Method (DEM) simulations is used to explore the dynamic behaviours of the materials in the mill. The results are then used to validate a novel approach of grindability assessment of pharmaceutical materials using the Scirocco disperser of the Malvern Mastersizer 2000 (Bonakdar et al. 2016). Based on this method the ease of grindability of the materials in the mill can be predicted.

Bonakdar, T., Ali, M., Dogbe, S., Ghadiri, M., and Tinke, A. (2016). A method for grindability testing using the scirocco disperser. *International Journal of Pharmaceutics,* 501: 65-74

507857 Bi-Directional Thermal Control of Twin Screw Granulation Process Via a Specialised Annular Heat Pipe

Abstract:

The continuous process of twin screw granulation forms a hotbed in pharmaceutical research as it is expected to succeed the current batch production of granules. In essence, the scale up of the material output is a function of production time, instead of equipment size (for batch granulators). This way, a â€œjust-in-time productionâ€ of the desired amount of material can be performed according to specific needs. In this work, a twin screw granulator was utilised in the granulation of both the placebo and active pharmaceutical formulations. Subsequently, the same equipment was used in granulating ceramic raw material powders for tile pressing. During processing, the spatial constraint in the barrel and screw compartments caused friction, resulting in excess heat generation. For the thermally-sensitive pharmaceutical formulations, this must be controlled to avoid denaturation of the granule products. A common solution is by using flowing cool water, but a good control has been proven to be difficult to achieve as different granulation zones generate different heat levels (kneading blocks > conveyer blocks). Heat pipe is a passive and highly effective thermal conductor and stabiliser. A carefully-designed annular heat pipe, filled with a small amount of methanol was successfully employed to stabilise the temperature in the barrel space, cooling the granulating material before it reached its critical temperature. For the ceramic powders, the methanol filling of the heat pipe was replaced by water and it was then heated up to provide an in-situ drying operation as the powders granulating. The in-situ drying in this instance, reduces energy consumption for the downstream drying processing as the product granules contain less moisture to be removed. The annular heat pipe, with its bi-directional operation of heating and cooling is a technically viable temperature controller for twin screw granulators. Compared to the established active cooling methods, the heat pipe provides a passive alternative, with only a discreet amount of fluid required to last its operational lifetime, thereby providing an economic thermal control solution.

Mixing Grains with Different Elongation in a Rotating Drums

Claudia Piacenza[1,2], *Marco Marconati*[1], *Colin Hare*[1], *Andrea C. Santomaso*[2], *Marco Ramaioli*[1,*]

[1] *Department of Chemical and Process Engineering, Faculty of Engineering and Physical Sciences, University of Surrey, Guildford GU2 7XH, UK*
[2] *APT Lab, Department of Industrial Engineering, University of Padova, Padova, Italy*

[*] *Email: m.ramaioli@surrey.ac.uk*

Abstract

Mixing grains with different properties is a remarkably challenging process, relevant to many industrial applications. Rotating drums have been used extensively as model systems to study granular media flow and mixing and segregation. Numerous studies considered the mixing of grains with different sizes, but only few studies considered shape and elongation, which have been already identified as important characteristics affecting packing and inducing segregation. In this contribution, the mixing of binary mixtures of grains having the same volume, but different elongations was studied experimentally. The mixing dynamics of a layered granular medium was characterized in a rotating drum, highlighting the impact of the grain shape, of the drum angular speed and of the drum filling ratio. A mixed or segregated state is reached very rapidly, but experiments are continued to verify that the state reached is a steady state. The experiments demonstrated that below a critical difference of elongation, the grains can be mixed effectively. Conversely, when grain elongations are very different, a central core is formed, rich in the more elongated grains. In this case only limited mixing can be achieved. These results can guide the formulation of mixtures of grains with different elongations.

1. Introduction

Mixing is a process operation aimed at obtaining a homogeneous system starting from two or more components. On the contrary, when there is a lack of homogeneity within the mixture segregation occurs. Segregation phenomena represent a relevant problem in the case of mixing solid particles having different physical properties such as particle size, density, shape or surface roughness. Differently from liquids and gases, solid particles are not thermal activated system [1] and so in order to obtain a certain degree of mixing they need to be mechanically activated. Typically three mixing mechanisms are largely recognized:

- dispersion, analogous to diffusion in fluids, it involves single particles that randomly move due to collision between each other;
- convection, it involves a portion of the bulk material, with a given direction and velocity;
- shear mixing: it is peculiar to granular solids and it is a combination of the two previous mechanisms.

Powder mixers are relatively simple machines of low capital cost. They are classified into two types: tumbling mixers and agitated mixers. Tumbling mixers are the simplest. A totally enclosed vessel is rotated about an axis causing the particles within the mixer to tumble over each other on the mixture surface. Tumbling vessels provide low-shear environments. Convective mixers instead are characterized by the action of impellers or paddles to move the powders around and generate a well-mixed product. Convective mixers provide more shear into the mixture and can be used with cohesive materials.

Rotating drums belong to the tumbling mixers. They are extensively used in the chemical and process industries. With respect to mixing processes, rotating drums are often used to study and understand mixing and segregation phenomena; as a result, many experimental and numerical studies were developed on granular behavior in rotating drums [1-3]. Different granular flow regimes can occur in the drum, each one with its specific flow behavior. In particular, different flow regimes can be assessed by using the dimensionless Froude number, $Fr = \omega^2 R/g$ [4], where ω [s^{-1}] is the drum rotation speed, R [m] the drum radius and g [m/s^2] the gravitational acceleration. With $10^{-4}<Fr<10^{-2}$ and filling level $\varphi>0.1$ rolling regime takes place. This motion is characterized by a uniform, steady flow of a particle layer on the surface (cascading or active layer) while the larger part of the bed (plug flow region) is moved upwards by solid rotation with the rotational speed of the wall. A uniform and good mixing is achievable with this type of motion, but when particle size distribution is very broad segregation can appear. The rolling bed is preferred, since is the simplest regime providing favorable conditions for heat transfer and high quality of the product even when mass flow rates are large.

When particles to be mixed have different properties such as different size, different density or different shape segregation typically occurs and particles tend to separate. [5, 6]. Difference in size is by far the most important cause of particle segregation. Particle size distribution influences also strongly bulk properties such as packing fraction and bulk modulus [7].

Different size segregation mechanisms can be identified:

- trajectory mechanism: according to the Stokes' law the velocity of fine particles is lower than that of big particles. This can cause segregation when small particles move through air;
- percolation mechanism: if a bulk material is disturbed, a rearrangement in the particle packing occurs and for fine particles it is easier to fall down in the inter-particle voids;
- inertial mechanism: in pouring and filling operations bigger particles tend to flow selectively to the bottom of the pile since they have a higher momentum;
- elutriation mechanism: in presence of air the very fine particles can remain in suspension while larger ones tends to settled. This can occur in charging storage vessels or hoppers.

A homogeneous binary granular mixture, with differences in size and density, in a two-dimensional rotating drum typically gives radial segregation. Considering a rotating drum operating in the rolling regime filled with particles of different sizes, bigger particles mainly tend to the edge of the system, while smaller particles accumulate to the centre of the drum. The phenomenon is fast and the rotational speed of the drum seems not to affect the phenomenon; the size ratio of the particles is, instead, the parameter that mainly affects size segregation phenomena. Most of the experimental investigations about segregation in the rotating drum involved spherical or nearly-spherical particles; the experimental studies of the flow of non-spherical particles are, in fact, rather limited [6,8]. The shape of the particles, however, strongly influences the motion in the system. In particular, the static and dynamic angles of repose of non-spherical particles are found to be very different from the ones measured for spherical particles suggesting that differences in particle mobility can affect the segregation process [9].

2. Materials and methods

2.1 Particles

Both spherical and non-spherical particles were used for the experiments. Their size (diameter d for spheres, major and minor axis for ellipsoids a and b respectively) , aspect ratio AR, projected area A_p, volume V, and true density ρ are reported in Table 1.

Table 1. Physical properties of the particles

Type	d [mm]	a [mm]	b [mm]	AR [-]	A_p [mm^2]	V [mm^3]	ρ [g/cm^3]
Spheres S	8	-	-	1	50.3	268	1.09
Ellipsoids E1	-	10.48	6.98	1.5	57.5	267	0.96
Ellipsoids E2	-	18.44	5.26	3.5	76.2	267	0.98

Spherical particles with size 8 mm were used for the experiments. The ellipsoids were 3D printed with *AR* equal to 1.5 and 3.5. They were printed with the Ultimaker 3D printer and made of polylactic acid (PLA). The software used to realize the 3D design of the particles used in this work is FreeCAD, which is an open-source parametric 3D modeller allowing to obtain STL files.

The true density of all the particles used for the experiments was measured by gas pycnometry (Multivolume Pycnometer 1305, Micromeritics). The measured density of the spheres was very close to the typical density of the nylon 66 (1.14 g/cm^3). The PLA ellipsoids had a density that was lower than the typical density of the PLA (1.24 g/cm^3). This is because the particles were printed with a set infill density of the 20%, so they where partially empty inside.

2.2 Tumbling mixer

The quasi-2D drum was made of two transparent polycarbonate disks with a diameter of 30 cm and a thickness of 6 mm; between them a wooden spacer was placed, with the same external diameter, an internal diameter of 25 cm and a thickness of about 9.1 mm. The spacer allowed to create an internal volume of finite thickness separating the two polycarbonate disks.

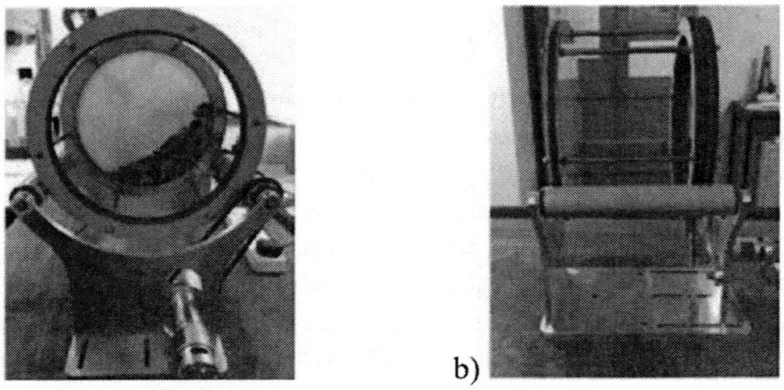

a) b)

Figure 1. a) Front view and b) side view of the rotating drum (the drum is on the left side of the rigid structure).

The internal lateral wall of the volume was covered with sandpaper in order to provide friction and avoid undesired slip of the particles. To keep the drum in vertical position and allow its rotation, three additional wooden spacers were used, at a distance of about 16.3 cm. This rigid structure could rotate at variable rotational speed thanks to a motor connected to a voltage supplier. Two rollers mounted on a rigid support made the rotation of the drum possible; a pulley transmitted the motion to the driving roller, while the second roller was passive. The rotational speed could be changed by varying the motor voltage. The structure could be removed from the rollers and by unscrewing the bolts, the drum could be easily opened so that to load/unload the particles.

2.3 Image acquisition

Frontal pictures of the drum were taken by a color camera and analyzed by the image processing software ImageJ [10]. To avoid the reflection of the camera and other objects on the front transparent disk, a squared LED panel was used which allowed to have a dominant diffused light illuminating the drum. The LED had a power of 8 W and a color temperature of 4000 K; it was placed behind the drum and opposite to the camera. A circular sheet of paper, which was attached to the drum between it and the LED, diffused the light properly. The camera was a Basler ace camera – model acA2000-165uc, with a maximum resolution of 2040 × 1086 pixels. The focus and the diaphragm of the camera could be adjusted manually.

2.4 Experimental plan

The attention was focused on mixtures of particles with different shapes, since segregation phenomena due to size had been widely investigated in the past. Three mixtures were considered, with two possible initial top-bottom segregated configurations. Filling level and rotational speed where changed also as detailed in Table 2.

Table 2. Experimental tests performed with differently shaped particles.

AR 1 [-]	*AR* 2 [-]	ω [rpm]	φ [-]	Init. Config.
1	1.5	4.22	25%	E1/S
1	1.5	4.22	25%	S/E1
1	1.5	4.22	47%	E1/S
1	1.5	4.22	47%	S/E1
1	1.5	11.51	25%	E1/S
1	1.5	11.51	47%	E1/S
1	3.5	4.22	25%	E2/S
1	3.5	4.22	25%	S/E2
1	3.5	4.22	47%	E2/S
1	3.5	11.51	25%	E2/S
1	3.5	11.51	25%	S/E2
1	3.5	11.51	47%	E2/S
1.5	3.5	4.22	25%	E2/E1
1.5	3.5	4.22	25%	E1/E2
1.5	3.5	4.22	47%	E2/E1
1.5	3.5	4.22	47%	E1/E2

*AR*1 and *AR*2 indicate the aspect ratios of the particles in the binary mixture. It is reminded that the spheres (*AR*=1) and the two types of ellipsoids (*AR*=1.5 and *AR*=3.5) had the same volume. The E1/S initial configuration indicates a segregated configuration with short ellipsoids above spheres (this was assumed as basis configuration); the S/E1 configuration is inverted with spheres above short ellipsoids. A similar convention is applied to the experiments considering mixtures of ellipsoids. All the experiments were performed with binary mixtures with a composition of 50% and 50% on a number (and hence also volume) basis. The filling level φ of the 25% was obtained with a total number of particles equal to 200; with 320 total particles the filling level of the drum was equal to 47%.

2.5 Image analysis procedure

Image analysis techniques were used to identify the particles on the basis of their different shape. The procedure was automated using customized ImageJ macros. Contrast and brightness of the images were first adjusted. Color images were then set to gray scale and segmented by applying a threshold. The operation was carried out twice to separate light and dark particles with different thresholds. Watershed segmentation was then used to separate touching objects. This last operation was more difficult for the particles at the border of the drum; in some case, in fact, the ellipsoids were erroneously identified as spheres. Also, the presence of the border of the drum complicated the watershed operation.

The centre of mass of each particle ($x_{m,i}$ and $y_{m,i}$) was estimated and the radial distance (d_i) from the centre of the drum (x_c, y_c) calculated as:

$$d_i=[(x_c-x_{m,i})^2+(y_c-y_{m,i})^2]^{0.5}$$

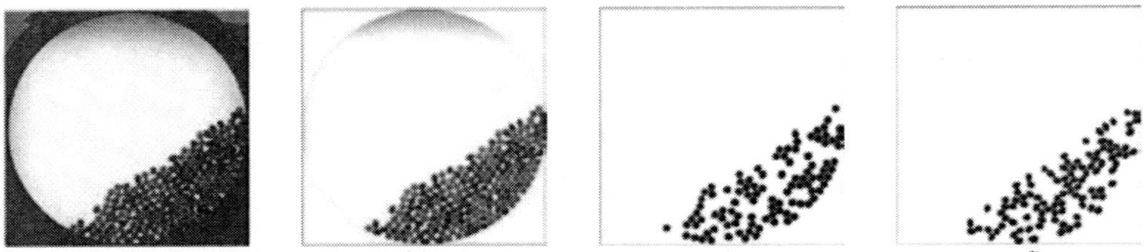

a) b) c) d)

Figure 2. Example of image processing steps. a) Original colour image; b) grayscale conversion and circle cut from the original picture; segmentation applied to c) light particles and d) dark particles.

The mean value of distance (d_m) for the spheres and for the ellipsoids was then calculated as:

$$d_m = \Sigma\, d_i/N$$

Only for very elongated ellipsoid (AR=3.5) the watershed operation failed resulting in over-segmentation errors. Single particles could not be identified unequivocally in this case; only one center of mass for all the elongated ellipsoid was estimated and the average distance (d_m) of the particles from the drum center calculated.

3. Results and discussion

Figure 3 shows two typical experiments, carried out at the lowest rotational speed and filling ratio, with ellipsoids (AR=1.5 at left and AR=3.5 at right) on top and spheres on bottom in the initial and final configuration. It is evident even by visual inspection that the final state presented some level of radial segregation and that a different final segregation intensity was reached in the two systems.

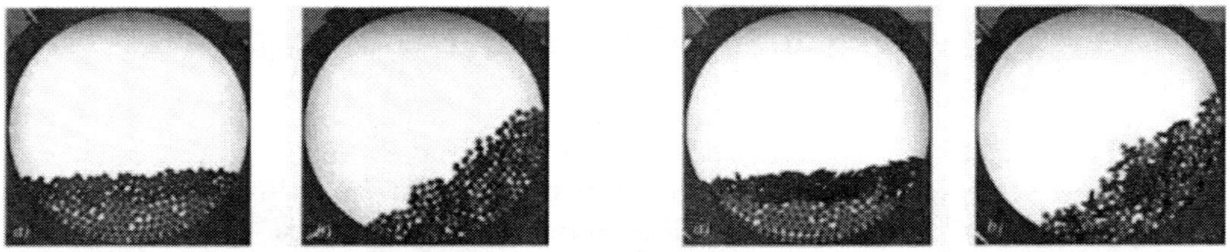

Figure 3. Drum at a) t=0 and b) t=30 min. On the left the initial configuration is E1/S (using ellipsoids with AR=1.5) and on the right the initial configuration is E2/S (AR=3.5). In both experiments ω=4.22 rpm and φ= 25%.

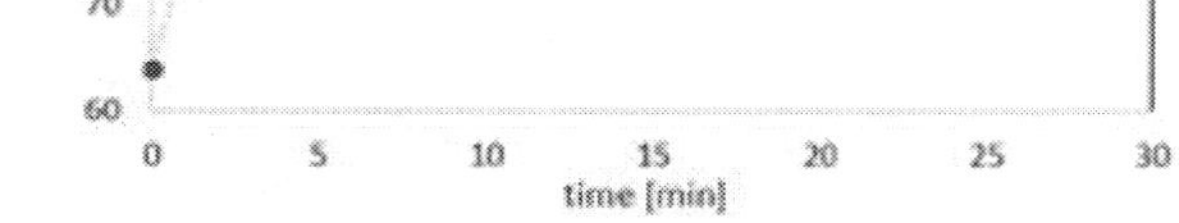

Figure 4. Evolution of the mean radial distance vs. time for the mixture of spheres and ellipsoids with AR=3.5. The initial configuration E2/S is compared to the initial configuration S/E2.

A more quantitative analysis was obtained by plotting the trend of the mean radial distance (d_m) of the two species at different time instants. As a typical example of the system evolution, the case of E2/S

is shown in Figure 4. The d_m reached a steady value after a short time. Dedicated experiments on a shorter time scale (2 minutes, not shown here) revealed that steady state value for d_m was reached after 30s for mixtures with ellipsoids E1 (*AR*=1.5) and after 60s for mixtures with E2 (*AR*=3.5).
At steady state, the radial distance was in both cases smaller for the ellipsoids, meaning that they mainly remained concentrated on top or at the center of the granular bed while the spheres mainly concentrated at the periphery of the drum (close to the drum wall). This tendency was much stronger for ellipsoids E2. So a slower segregation dynamics compensated by a larger segregation intensity was observed with the more elongated particles.
As Figure 4 shows, inverting the initial configuration from E/S to S/E did not affect the dynamics and the final result in all the experiments performed with these particles. Initial configuration resulted to have no effect on the segregation propensity of the mixtures analyzed.

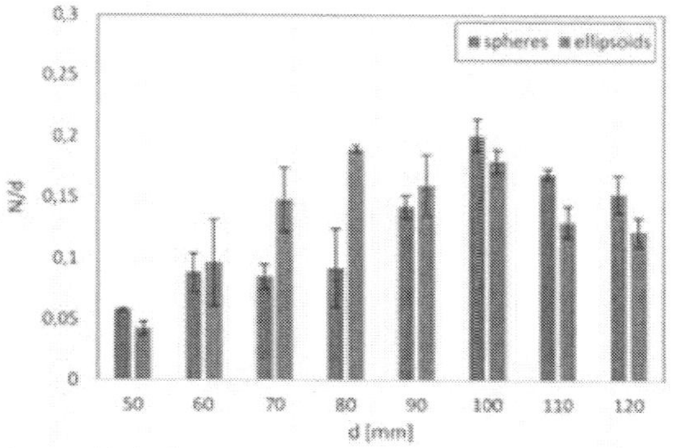

Figure 5. Distribution of the radial distance of spheres and ellipsoids with AR=1.5, obtained by averaging the results of three repetitions. Other conditions are t=30 min, ω=4.22 rpm, φ= 25%.

A more in-depth analysis can be carried out considering the distribution of the radial distance of the spheres and ellipsoids separately. The radial distance from the centre of the drum (*d*) was divided in classes of 10 mm and the number of each type of particles (*N*) in each class was computed. To obtain the distribution, *N* was normalized on the radial distance as *N/d* and plotted versus *d*. In the case of completely homogeneous mixtures and assuming that the medium takes the shape of a circular sector, *N/d* should be constant and equal for the two types of particles in each class. The peak of the distribution of the ellipsoids was found at lower distance than the peak of the distribution of the spheres. Moreover, at lower distance, the number of ellipsoids was higher than the number of spheres, especially for distances between 70 and 90 mm; at higher distance (closer to the drum wall), instead, the number of spheres was higher than the number of ellipsoids, thus confirming the previous analysis based on d_m.

Also, binary mixtures made of the two types of ellipsoids E1 and E2 (*AR*=1.5 and *AR*=3.5 respectively) were taken into account at the same filling levels of the previous experimental tests and at the lower rotational speed. Stable segregation was again observed with intermediate results with respect to the two previous cases (S+E1 and S+E2). So as a general comment on these experiments, stable segregation was always observed when mixing particles having same volume but different *AR*. The intensity of segregation increased with increasing the difference in *AR* between the particles. In particular particles with larger *AR* tended to accumulate at the center of the granular bed, while particle with lower *AR* accumulated at the periphery. Segregation occurred in the surface active layer where the more elongated particles, because of their reduced propensity to flow (particle rolling was prevented by their elongated shape), were systematically overtaken by the less elongated (and faster) particles. In this way a spontaneous separation of particles occurred in the active layer, with those having a lower *AR* arriving first at the drum wall, while those having an higher *AR* remaining trapped in the core.

3.1 Effect of filling level

The comparison of the behavior of the mixtures at the two filling levels can be carried out by normalizing the mean radial distance d_m of the two types of particle on the mean distance of all the particles d_c, calculated as $(d_{m,\ spheres}+d_{m,\ ellips.})/2$. The trends of d_m/d_c are shown in Figure 6

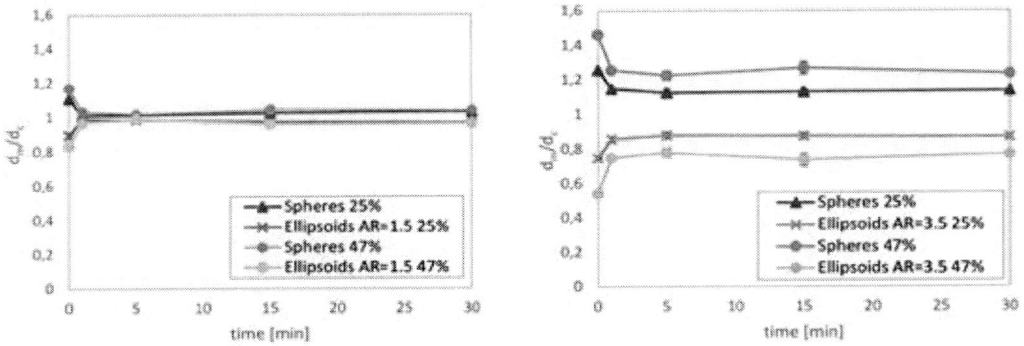

Figure 6. Trend of d_m/d_c vs. time for the mixture of spheres and ellipsoids with AR=1.5 (left) and AR=3.5 (right) at two different filling levels.

In the case of mixture of spheres and ellipsoids with AR=1.5, the curves at different filling levels overlap, showing that the behaviour of the mixture was not affected by the filling level of the drum. In the case of the other two mixtures (S+E2 and E1+E2, not shown), the curves did not overlap , especially for the case S+E2 shown in Figure 6 (right). An explanation can be found in terms of bed geometry. When increasing the filling level (below 50%) the length of the surface active layer increased as well. This gave more chances for particles to separate, in particular when the difference in AR was large enough (S+E2 and E1+E2). However for the system S+E1 where the difference in AR was the smallest, the difference in active layer length in the two cases was not sufficient to create observable differences in the segregation pattern.

3.2 Effect of rotational speed

In addition to experiments carried out at 4.2 rpm, tests were repeated at 11.5 rpm with spheres and ellipsoids (25% filling level). Figure 7 shows that rotational speed did not affect the segregation patterns, at least in the range of the analyzed speeds.

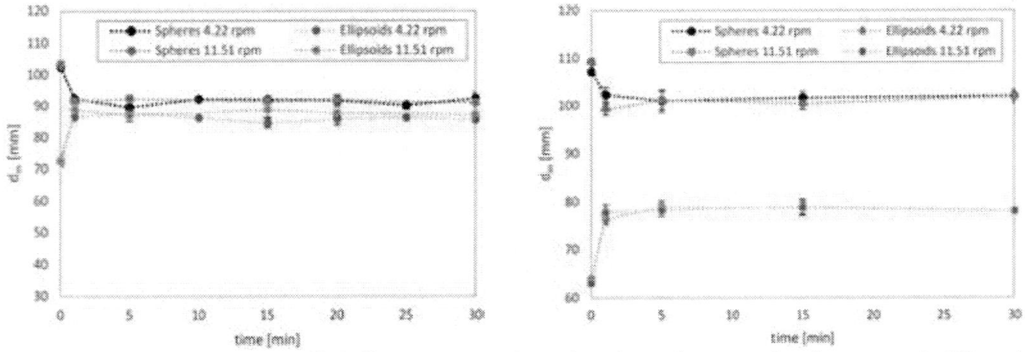

Figure 7. Trend of the mean radial distance vs. time for the mixture of spheres and ellipsoids with AR=1.5 (left) and AR=3.5 (right) at two different rotational speeds.

Conclusions

From the experimental tests described in the previous paragraphs, it was found that binary mixtures of particles with different shape but with the same volume do not mix uniformly and segregation was always observed. In particular, the particles with lower aspect ratio tended to stay closer to the drum walls, while the more elongated particles were found preferentially at the centre of the granular bed. The most significant case was represented by the mixture of spheres and ellipsoids with AR=3.5; for this mixture, in fact, the segregation was more evident and the non-spherical particles clearly accumulated in the centre of the granular bed. The same final state was obtained regardless of whether the longer particles were initially located to the top or the the bottom of the mixture. A proportionality between AR difference and segregation intensity was observed. Some variables such as rotational speed and initial bed configuration were found to not affect the segregation dynamics and intensity. Only filling level (in addition to AR) was found to affect segregation intensity. A mechanistic explanation was given in terms of differences in particle flowability occurring in the surface active layer (confirmed also by an analysis of the angles of repose in the drum). In this specific case therefore percolation and inertial effects (typical of size segregation) were not the main driving mechanisms for the observed segregation (the particle indeed had the same volume and similar mass so that sterical and inertial phenomena were minimized). Segregation mainly occurred because of differences in particle velocity in the active layer due to a reduced mobility of particles with larger AR.

References

1. GDR MiDi (2004), "On dense granular flows", The European Physical Journal E, 14, 4, 341–365.
2. Jarray A., Magnanimo V., Ramaioli M., Luding S. (2017) Scaling of wet granular flows in a rotating drum, EPJ Web of Conferences, 140,03078.
3. Volpato S., Canu, P., Santomaso A.C. (2017), "Simulation of free surface granular flows in tumblers", Advanced Powder Technology 28(3), 1028-1037.
4. Mellmann J. (2001), "The transverse motion of solids in rotating cylinders – forms of motion and transition behaviour", Powder Technology, 118, pp 251-270.
5. Windows-Yule C.R.K., Scheper B.J., van der Horn A.J., Hainsworth N., Saunders J., Parker D.J., Thornton A.R. (2016), "Understanding and exploiting competing segregation mechanisms in horizontally rotated granular media", New Journal of Physics, 18, 2, 023013, 2016.
6. Santomaso, A.C., Olivi, M., Canu, P. (2004), "Mechanisms of mixing of granular materials in drum mixers under rolling regime", Chemical Engineering Science, 59 (16), 3269-3280.
7. Kumar N., Magnanimo V., Ramaioli M., Luding S. (2015), "Tuning the bulk properties of bidisperse granular mixtures by small amount of fines", Powder Technology, 293, 94-112.
8. Ramaioli M., Pournin L., Liebling Th. M., (2005), "Numerical and experimental investigation of alignment and segregation of vibrated granular media composed of rods and spheres", Proceedings of the 5th International Conference on Micromechanics of Granular Media, 2, 1359-1363.
9. Santomaso, A.C., Canu, P. (2007), "Single particles properties vs. bulk flowability", Proceedings of 5th International Conference for Conveying and Handling of Particulate Solids.
10. Schneider, C. A., Rasband, W. S., Eliceiri, K. W. (2012), "NIH Image to ImageJ: 25 years of image analysis", Nature methods 9(7): 671-675.

507441 Generation of Particles with a Special Morphology By Desublimation of Copper Phthalocyanine

Abstract:

The morphology of submicron particles becomes more and more important concerning new functional materials. For example the special electric properties of particles can be used in electronic devices [1]. Moreover the electric, optical or catalytic properties of particles can be optimised by adjusting the morphology of the particles. This may imply a risk for the health but makes it possible to realize specific product characteristics. The morphology of the particles should therefore be considered e.g. for filter tests and for the developing and testing of sensor systems.

To explore the characterization of the particle morphology the DFG-AiF cluster project "multi-parametric characterization of particle-based functional materials using innovative online measurement systems" (MPaC) includes the development of online sensor systems for the size and structure of the particles as well as the development of generation methods for particles with a special morphology. The production of spherical particles can be achieved with different commercial aerosol generators. Several generation techniques for special particles are described in literature [2] [3] [4]. To test the sensor systems with particles of different shapes a new type of aerosol generator was developed. The aim for the new generator is to produce the particles in the gaseous phase and provide the particles directly for sensor or filter testing.

The developed generator uses the desublimation of copper phthalocyanine (organic pigment) to structure the surface of titanium dioxide. The bulk material copper phthalocyanine is dispersed into a nitrogen gas stream and transported to a tube furnace. Here the bulk material sublimates and the gas stream carries the material to a laval nozzle where the adiabatic expansion takes place. This leads to a rapid decrease of the temperature behind the laval nozzle and an oversaturation in the gaseous phase. In addition to the decrease of the temperature core particles of titanium dioxide were initiated behind the laval nozzle. The copper phthalocyanine desublimates on the core particles and causes a structured surface. The morphology of the generated particles is characterized by SEM. The particles show a punctual coating of the titanium dioxide particles with copper phthalocyanine. The size of the titanium dioxide particles is about 500 nm and the punctual coating of copper phthalocyanine measures a few nanometres.

The generation method and the design of the generator will be presented and the morphology of the generated particles will be described.

[1] K. Deppert, M. H. Magnusson, S. B. Carlsson, B. Gustafson, T. Junno, . L. Montelius, B. J. Ohlsson, L. Samuelson und L. E. Wernersson, „Future electronic devices by aerosol techniques," *Journal of Aerosol Science,* Nr. 31, 2000.

[2] J. Raula, A. Kuivanen, A. Lähde, H. Jiang, M. Antopolsky, J. Kansikas und E. . I. Kauppinen, „Synthesis of image-leucine nanoparticles via physical vapor deposition at varying saturation conditions," Journal of Aerosol Science, Nr. 38, 2007.

[3] M. Heurlin, M. H. Martin, D. Lindgren, M. Ek, R. L. Wallenberg, K. Deppert und L. Samuelson, „Continuous gas-phase synthesis of nanowires with tunable properties," Nature, Nr. 492, 2012.

[4] K. Nielsch, K. Deppert, M. H. Magnusson, G. Dumpich, E. Kruis und H. Fissan, „Formation of ultrafine particles from powders of compounds with incongruent sublimation behavior," Journal of Aerosol Science, Nr. 28, 1997.

516184 Rubbery Milling of Seed Endosperms for Improved Sustainability by Natural Functionality Preservation

Abstract:

Increasing demand for functionalised goods addressing individual needs requires a profound understanding of process-induced alterations. Development of models and process-structure-property relationships are essential to control the resulting quality and properties of processed food. A wide range of food systems contain solid or semi-solid particulates as major structure building blocks. Generation and tailoring of such particulates strongly depends on their size distribution, shape, morphology, phase behavior (amorphous/crystalline), interfacial properties and related inter-particle force interactions.

A novel processing technology was developed widening the frame of possible alterations, aiming towards tailoring such particle properties generating novel food structures or preserving techno-functional characteristics of the raw material. The technology employs targeted adjustments of the initial moisture content in combination with milling temperature. The resulting transition from the glassy to the rubbery state generates different material properties and processing behaviour. The technology is highly versatile, owing to the universality of moisture induced state transitions.

Seeds from locus bean gum or guar gum were wetted with excess water in order to adjust a rubbery state of the seed endosperm material. In this state milling within a novel type of impact mill with superimposed drying and sifting steps causes more gentle material disintegration compared to conventional dry milling. This preserves the high molecular weight fraction of e.g. galactomannans (micro-scale structure) from degrada-tion and allows at the same time the adjustment of a micro-porous morphology (macro-scale structure) improving wetting, water uptake and dispersing characteristics of the resulting thickener powders. Such structure preservation /modification increased the thickening characteristics extraordinarily as will be demonstrated.

From this prominent rubbery milling example we have entered in related systematic work addressing a wider range of seed endosperms with various â€œnature-synthesizedâ€ func-tionnalities to be further addressed.

/1/ Pollard, M., Eder, B., Fischer, P., and Windhab, E. J. (2010) "Characterization of galacto-mannans isolated from legume endosperms of Caesalpinioideae and Faboideae subfamilies by multidetection aqueous SEC." *Carbohydrate Polymers*, 79(1), 70-84.

/2/ Illmann, S., Stoob, M., Sieber, C., Konz, A.K. and Windhab, E.J. (2012) "New Thickening Quality of Galactomannan Polysaccharides by Tailored Impact Milling" *XVIth Int. Congress on Rheology*, Lisbon, Portugal, August 5-10

503422 Particle Technology Education at Purdue University

Abstract:

Degree programs and coursework at Purdue University focused on particle technology will be presented. Additional information may be found at: https://engineering.purdue.edu/CP3

506453 Teaching Particle Technology in Portugal â€“ University of Coimbra

Abstract:

Education in Particle Technology started in Portugal in the 80â€™s (last years of the decade). The Chemical Engineering degree of the University of Coimbra was the first engineering degree, in the country, to include this topic in the curriculum. Particle Technology was recognized by us, at that time, essential for the education of chemical engineers, in parallel to education in the traditional areas of Chemical Engineering which were still focused on monophasic systems.

Since then, Particle Technology has always been present in the Chemical Engineering curriculum of the University of Coimbra, in different configurations. During the first years particle characterization was combined with design of operations involving particulate systems (grinding and pneumatic and hydraulic transport). From 2005 to 2010, with the curriculum reformulation in relation with the Bologna Process, the two topics started to be addressed in different courses: one course on Materials Characterization and the process topic in a course entitled Structuring and Stabilization Processes, dealing with colloids and emulsions, grinding, agglomeration and spray drying and, for the first time, an integrated modelling strategy for particulate processes based on the population balance formulation was introduced in this course. The two courses were compulsory, at this time.

From the school year 2011/2012 till present, the configuration of this area in the Chemical Engineering degree of the University of Coimbra changed again, even if not deeply (see https://apps.uc.pt/courses/EN/course/361). The course on materials characterization, which includes general notions on particle characterization is kept as an autonomous course included in the 1st cycle of the Chemical Engineering degree of the University of Coimbra. A second course on design of processes involving particulate systems now exists in the 4th year of the MSc degree (1st year of the 2nd cycle), entitled Disperse Systems Technologies, which is an optional course, and which deals with all types of disperse systems, i.e gas, liquid or solids as disperse phase. This course starts with a review of the mechanisms/interactions occurring at the interface with the disperse phase, goes on, again, with a chapter on population balances modelling, and proceeds with strategies for disperse systems stabilization, linking this subject with formulation concepts, ending with the design of some processes involving particulate systems, such as colloids and emulsions, crystallization, grinding, agglomeration and spray drying. This course is also offered to other engineering degrees, such as Materials Engineering and Biomedical Engineering.

Also at the PhD level a course on Disperse Systems Technologies can also be offered during the formal lectures period of the PhD program on Chemical Engineering (or others), depending on the students background and interests.

Since the introduction of the Particle Technology topic in the curriculum of the Chemical Engineering degree of the University of Coimbra, the feedback from employers has been very positive, and the competences in this field acquired by our students have been quite valued by the job market. Till today, the Chemical Engineering degree of the University of Coimbra is still the one, in Portugal, with a stronger emphasis on this topic in the curriculum. In our curriculum, the topic of Particle Technology interacts directly with another course on Product Design, also in the 4th year of the MSc degree and, of course, also with the Project Design course (5th year of the MSc degree), besides other more traditional courses in a Chemical Engineering degree such as in Reaction Engineering and with courses related to environmental technologies.

At present, we have started developing multimedia tools for the teaching/learning of this topic, especially in the field of particle characterization, to be made available on-line, as is the case of the interactive video on nanoparticles characterization (http://labvirtual.eq.uc.pt/siteJoomla/index.php?option=com_content&task=view&id=277&Itemid=470, in Portuguese). Also, the European Federation of Chemical Engineering (EFCE) is considering investing in this type of tools for the field of particle technology, as the result of a collaboration between the Working Groups of Characterization of Particulate Systems and Education.

Acknowledgement

The authors would like to acknowledge the financial support of Pest/C/EQB/UI0102/2013 from FCT/MCTES (PIDDAC), co-financed by the European Regional Development Fund (ERDF) through the program COMPETE (POFC).

506629 Highlights of Particle Technology Teaching in Singapore

Abstract:

Our approach to education in particle technology is to incorporate elements of our research in this field, relevant industrial case studies and apply a Technology Enhanced Learning (TEL) approach for implementing pedagogical interventions to enhance the learning experiences of students.

For supercritical CO_2 processing, case studies in extraction and drying are used to teach different chapters in the curriculum such as:

- Particle characterization
  - The bulk density, size, shape of real industry samples are given to students for analysis using software such as ImageJ
- Flow through packed bed
  - The pressure drop and scale-up of packed bed design for supercritical CO_2 flow through packed bed are used as a worked example and assignment problem to help students appreciate the role of density and viscosity in the pressure drop in a packed bed
- Drying
  - Supercritical CO_2 as an alternative drying media as compared to hot air or superheated steam. Case studies and examples provided

For gasification and chemical looping combustion applications, case studies in the treatment of wastes and carbon capture are similarly used as part of the syllabus:

- Particle characterization
  - The effect of polydispersity on various fluidization characteristics (entrainment, transport dis-engaging height, clusters, segregation, etc)
  - The effect of agglomeration
- Reliability of correlations and simulations
  - Correlations in literature give predictions orders-of-magnitude apart
- Scale-up
  - Rule-of-thumbs in scaling up the various regimes

A design software has been developed within the paradigm of Technology Enhanced Learning (TEL) to facilitate learning via a repeated practice approach. The software generates detailed solution steps to typical engineering design problems encountered within this module. Students can compare the complete solutions generated with their own hand calculations, and thereby apply a repeated practice approach towards their learning of engineering design calculations. The software has been successful in enhancing the performance of students in the final examination.

507116 Particle Technology Course at the University of Salerno

Abstract:

The Course of Particle technology at the University of Salerno is a 6 EC, that is a 60 hours course including lectures, as well as classroom and laboratory practical. The course dedicated to Master student in Chemical Engineering. It a compulsory course within the â€œinnovative process and nanotechnologies curriculumâ€ can be optioned by any mechanical or chemical engineering student at the master level.

The course includes basic particle size and shape characterization concepts, fundamentals on interparticle forces, fundamentals of powder mechanics and silo discharge, fundamentals on-fluid particle interactions and fluidization.

The main learning outcome is the ability of the student to understand the phenomena occurring in processes involving particulate solids and to carry out basic design operation on storage and fluidized bed units.

507382 Professional Master of Engineering Degree in Particle Technology

Abstract:

Particles are ubiquitous in the process industries yet nearly absent in the undergraduate education of chemical engineers. This poster descibes a professional Master of Engineering degree program at University of Delaware (the MEPT) that closes that gap by exclusively focusing on particulate matter, its processing, and its incorporation into products. The program can be completed in a single year although we welcome already-employed students who wish to take it part time. Now in our fourth year, we will describe our program and our experiences including guiding distance learners successfully through the program.

In its 1-year realization, a core curriculum of 18 credits is augmented by 6 credits of electives which the student may use to tailor the program to their specific interests. The program is capped by a 10-week, 6-credit summer internship. The 12 core credits taken in the fall teach the fundamentals of particle transport, particle properties & characterization, particle formation and transformation kinetics and the mathematics of particulate systems. The 6 core credits taken in the spring teach the student to apply the fundamentals to both (a) equipment design (in a unit operations course) and (b) product and process design (in a product engineering course).

The presenters of this poster are the program directors, both with long industrial careers preceding this venture into education. In addition, we have developed a strong network of industrial partners who share a wealth of additional experience and expertise with our students both as adjunct professors and as internship mentors. We have successfully placed international students in internships at US companies via the curriculum practical training program (CPT) who have gone on to practice in industry after graduation under the optional practical training program (OPT).

507780 Recent Developments in Particle Technology at the Universidad Tecnica Federico Santa Maria

Abstract:

The Department of Mechanical Engineering of the Universidad Técnica Federico Santa María formed in 2003 a Research Center for Transporting Granular Materials (CITRAM) to support educational activities and experimental research, as an extension for those ME students taking the regular course on Bulk Solids Handling and those developing a thesis in this area.

Over the years, many ME students have studied, investigated and modeled different topics including: pneumatic conveying of solid particles of different nature, size and density; identification of flow patterns developed in the pipeline, minimum conveying velocity and pressure drop; abrasive wear using a novel rubber-wheel tester; segregation of ores in stockpiles and silos; flow of granular materials in hoppers and chutes; heat transfer of granular materials; among others. The aim of this poster is to present and share the results of this work.

508520 Using Perusall to Enhance Student Learning of Particle Technology at Graz University of Technology

Abstract:

Perusall is a digital platform that allows students to collaboratively annotate textbooks, papers, and other similar classroom reading assignments. In order to enhance student learning, Perusall was introduced to help teach Particle Technology at Graz University of Technology (TU Graz, Austria). At Graz University of Technology, two Particle Technology courses exist in sequence; the first course is focused on fundamentals topics, whereas the second course is focused more on applications and separations topics. Approximately 60-75 students are registered for each course. For both courses, Perusall was used to assign reading materials, which were primarily journal articles that complemented the course topics. Although these assignments were non-compulsory due to national requirements, students were asked to use Perusall to read and annotate the course readings. Student participation and input regarding this digital approach to learning was recorded. This poster will highlight how Perusall was implemented in both Particle Technology courses as a tool to enhance student learning.

508524 A Graduate Course in Fluidization and Gas-Solid Flow Systems

Abstract:

The objective of this course is: a) to provide students with a fundamental understanding and design of Fluidization and Fluid- Particle flow systems, b) to provide students with an understanding of Computational Fluid Dynamics (CFD) and the kinetic theory approach to fluid-particle flows c) to provide students with the knowledge required to solve flow, heat and mass transfer problems in processes which involve particles and d) to provide students with sufficient knowledge to read and understand technical articles in areas related to fluid-particle processes. The course is designed for Master and PhD students in Chemical, Mechanical and aerospace engineering. The course includes: Fluidization concepts and regimes including bubbling, turbulent and circulating beds; Particle characterization, diagnostics and size reduction and agglomeration/granulation; Concept and design of Standpipes, non- mechanical valves, gas distributors, cyclones and hoppers; CFD approach to fluid-particle flow systems and fluidization including kinetic theory and dispersed phase modeling; Computational transport phenomena of fluid-particle systems including method of moments; and examples from energy, chemical and pharmaceutical industries. In addition the students will visit PSRI to see large-scale fluidized beds in operation.

The students are required to finish a team project. The teams consist of 3-4 students working on a project which will count as their final exam. The projects will primarily involve background research (about 80%) but some of the investigation (about 20%) should involve taking an additional step from what is currently known. This additional step can take the form of a research proposal, new theory, new insight based on current information, new calculations, new simulation and/or new data. The results of the projects will be communicated to the class and to the Course Instructors via a written report of 10-20 pages, and an oral presentation involving the professors and all of the group members.

508742 Education of Fluid-Solid Multiphase Flow at Department of Mechanical Engineering, Osaka University

Abstract:

In mechanical engineering field, education curriculum in undergraduate course concentrates on the basic subjects of mechanical engineering such as mathematics, mechanics, material mechanics, fluid mechanics, thermodynamics, control engineering, design, etc. Concerning the course relating to powder technology, we have a class of "Fluid-Solid Multiphase Flow" in the curriculum of graduate school of engineering.

The contents of "fluid-solid two-phase flow" is as follows;

1 Introduction: examples of multiphase flow

2 Definitions and fundamental concepts

3 Particle response time and Stokes number

4 Fluid force acting on particle

5 Fluid force acting on particle in dense system

6 Particle-wall and particle-particle interaction

7 Heat and mass transfer

8 Continuous phase flow equations

9 Modeling of fluid-solid multiphase flow

10 Discrete particle modeling and simulation

11 Numerical simulation method for continuous phase flow

This course includes basic concepts of powder technology, dynamics of particles in fluid flow, derivation of basic equations for locally-volume-averaged fluid flow, discrete particle modeling including DEM (Discrete Element Method), and basics of numerical analysis and modeling.

509420 Bulk Solids Handling Education at the KSU Bulk Solids Innovation Center

Abstract:

Kansas State University Bulk Solids Innovation Center (BSIC) was started in 2015 with university and industry partnership and is located in Salina, Kansas. This center is dedicated to bulk solids handling research, education and outreach and equipped with world class pneumatic conveying system (both dense and dilute) full scale test lab and other material testing equipment. The vision of this center is to become valued resources for companies who produce or handle bulk solids during their operations. Currently, this center offers short courses in various subject lines related to bulk solids handling such as storage and flow, pneumatic conveying systems, discrete element method, dust explosion and control and mixing and segregation. Duration for these short courses varies from two to three and half days and offered as live course with class room teaching and practical training. The fundamental course on bulk solids handling is offered as distance education online format. This center is also in the process of starting two year online (distance education) master degree program named professional master of technology specialized in bulk solids handling and one year graduate certificate program (online) in bulk solids handling.

Particle Measurement in High-Temperature Gas based on Light Scattering

Li-feng LU[1,3], Xiao-lin WU [1,3], Zhong-li JI [1,2], Ming-xing WANG[1,3]

[1] Department of Chemical Engineering, China University of Petroleum, Beijing, P. R. China

[2] Department of Mechanical and Transportation Engineering, China University of Petroleum, Beijing, P. R. China

[3] Beijing Key Laboratory of Process Fluid Filtration and Separation, China University of Petroleum, Beijing, P. R. China

Abstract: In order to achieve the particle measurement at 650 °C, a single particle measurement method and device suitable for high-temperature gas were developed. The influence of gas temperature on optical sensor of optical particle counter (OPC) was investigated in detail based on geometric optics theory and heat transfer theory. The thermal expansion of aerosol tube along the airflow direction was caused by the heat conduction between the tube and the gas, resulting in the position of the light apertures to shift to obscure the partial light beam. The invar material can be adopted to eliminate the above mentioned disadvantages. According to the principle of gas state equation, the optical parameters (refractive index) of the gas at different temperatures change with the change of its state parameters. Based on the geometrical optics theory and the geometric mathematical relationships, the theoretical and experimental dynamic models of the optical measuring volume with the change of the gas temperature were established. The established models were then used to analyze the influence of gas temperature on the optical measuring volume. The results showed that the influence of the temperature on the optical system is negligible. Meanwhile, particle detection experiments were conducted at different temperature to achieve a good performance of OPC.

Keywords: High-temperature gas; Optical measuring volume; Dynamic model

1 Introduction

With the continuous improvement of environmental protection and the safety requirements for industrial production, the technology of high-temperature purification is drawing more and more attention, especially in the fields of clean coal power generation, coal gasification, petrochemical industry and metallurgy [1]. For example, nuclear power plants often produce a large number of low radioactive solid wastes, and the combustible wastes usually account for a considerable proportion. The incineration and pyrolysis are effective treatment techniques for the combustible waste, during which the temperature of flue gas is 600~800 °C. With the aim to purify the flue gas, the German Karlsruhe Nuclear Research Institute, the French nuclear power station and the Japanese nuclear power station have installed the silicon carbide ceramic filter tube, aluminum silicate fiber and alumina ceramic filter tube at gas temperatures above 650 °C, respectively, and investigated their properties properties [2-4]. These filters have a direct influence on the purification quality of gas and the particle concentration in the downstream. In an industrial cracker system, the particle concentration is required to reach 0.014 mg/m^3 after the filters; the Integrated Gasification Combined Cycle (IGCC) requires the particle size to be smaller than 1 um with a particle concentration of less than 2.2 mg/m^3 and in the range of 1~10 um with less than 5.8 mg/m^3, and to remove the dust size of larger than 10 um [5]. In a Shell Coal Gasification Process (SCGP), the particle concentration is required to be less than 20 mg/m^3 [6]. In order to detect the concentration and size of the particles in pipes in real time for calculating the separation efficiency of evaluating the purification performance of the filtration system, the particle detection technology is needed to be improved [7].

The widely-used particle detection devices in industry are mainly based on the principle of light scattering [8-10], and the particle size and concentration are obtained through an inversion of the scattering light intensity and particle size, which are suitable for the conditions of below 200 °C. The method of first cooling and then detecting can be adopted for the particle detection under higher-temperature conditions, but droplet condensation and sublimation will inevitably occur and result in an incorrect particle samples.

With the aim to achieve the particle detection at 650 °C, an optimization was conducted on the existing particle detector (OPC) in this paper. And based on geometric optics theory [11] and heat transfer theory, a detailed research was performed on the influence of gas temperature on OPC optical sensor, including the dynamic change of the aerosol tube and the optical measuring volume (OMV). Finally, the experiments were conducted on the on-line particles measurement of different temperatures.

2 Experimental setup and method

Figure 1 gives the experimental setup for detecting particles in high-temperature gas, mainly including the heating unit, optical detection unit and gas purification and processing unit.

The heating unit consists of the vibration particle feeder and the air heater. The gas brings the particle out of the feeder, and is heated up to 1000 °C by the heater, which is monitored quickly by a PID temperature control instrument with easy operation.

The optical detection unit is used to achieve the experimental measurement of the characteristics of the OPC optical system and the on-line detection of particles, including the optical sensor and the OPC control host. The optical sensor is shown in Figure 2, which is composed of an aerosol tube and optical system. Aiming to avoid the influence of high-temperature gas on optical system, the aerosol tube (shown in Figure 3) has been developed, and is in a distance of 15 mm to the optical lens. In addition, the insulation coating with lower thermal conductivity (-80-1700 °C) is adopted to treat the outer wall of the aerosol tube. A K-type micro thermocouple was bent and installed into the light aperture to monitor the gas temperature at the center of the aerosol tube in real time without affecting the airflow and optical paths. Figure 4 shows the optical system of the sensor, which is composed of an incident light path and an observed scattering light path. Both the paths are arranged in 90° with the same optical lens and elements [12-14], and the intersected area is the center of the aerosol tube to form an OMV [15]. The intersected area is enough small and allows particles to pass through it one by one [16, 17]. The OPC host consists of the light source system (wavelength: 300-700 nm) [18], the photoelectric detection system and the signal analysis and processing system. In the particle detection, particle is illuminated by the incident light when passing through the OMV, and the scattering light is collected and transmitted to the photoelectric detector to achieve the conversion of a pulse voltage signal from an optical signal. Finally, the tested particle size is calculated based on the calibration curve of particle size and voltage amplitude [14, 19].

The gas purification and processing unit is composed of a vacuum pump, a flow controller, a dust collector and a heat exchanger. The high-temperature gas form the aerosol tube is completely cooled during the heat exchanger, the flow rate of which is adjusted by the flow controller. In order to avoid the damage of particles on the pump, the dust is captured and collected by a filter in the dust collector.

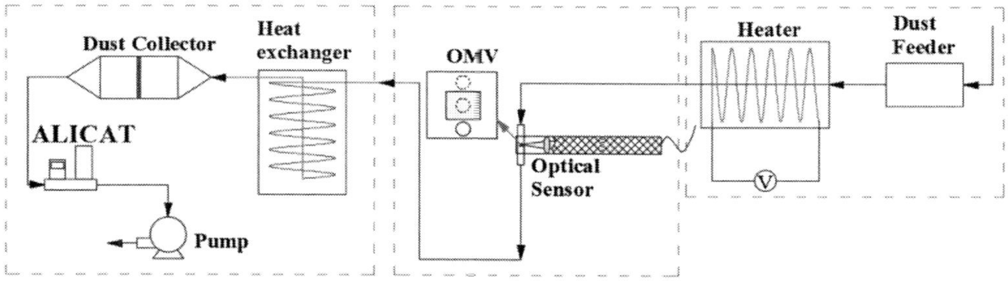

Figure 1. Experimental setup for detecting particles in high-temperature gas

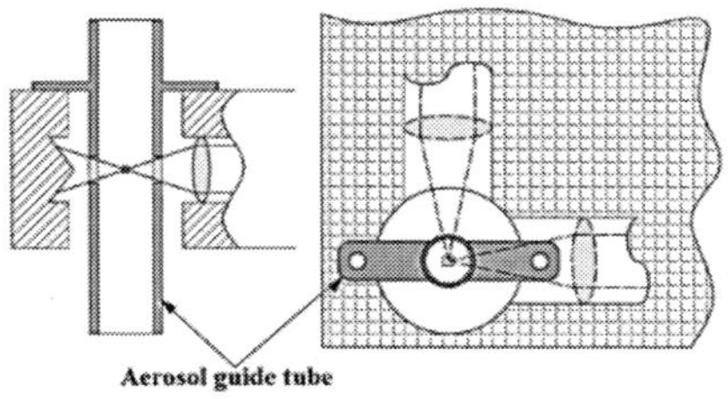

Figure 2. Optical sensor

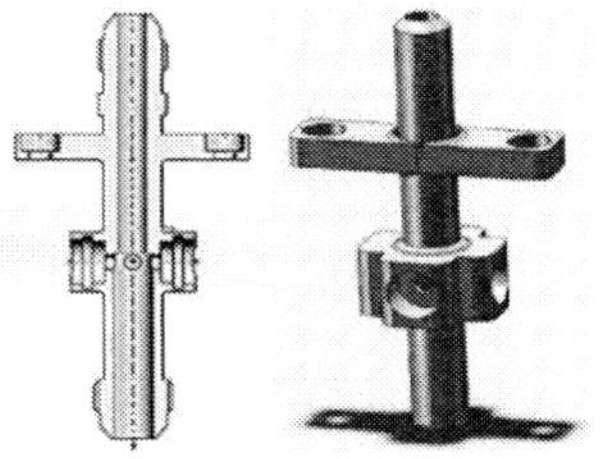

Figure 3. Aerosol tube for high-temperature gas

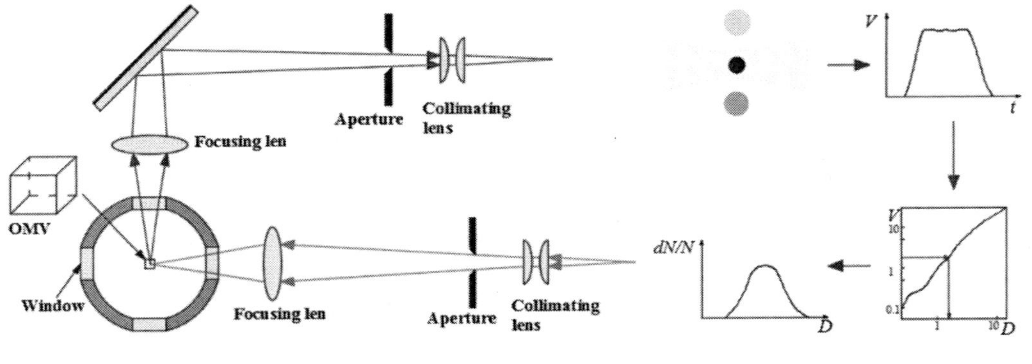

Figure 4. Optical system and principle of OPC

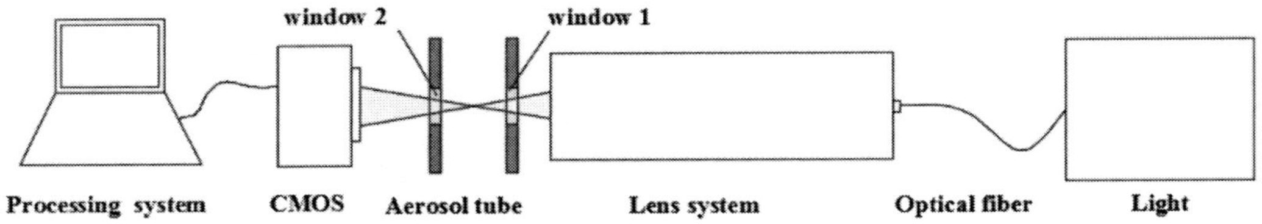

Figure 5. Evaluation and analysis device for light beam

3 Analysis and discussion

3.1 Influence of temperature on aerosol tube

The aerosol tube shown in Figure 3 has four apertures along the center axis, in which the glasses with a thickness of 2 mm were installed, as shown in Figure 4. When the gas flows in the tube, it can reduce the damage of high-temperature gas to the optical system elements. But the heat resistance of the tube has a direct influence on the imaging of the optical system at the center of the aerosol tube.

Figure 5 gives an evaluation and analysis device for light beam, including the light source system, the light path system, the aerosol tube and the light beam analyzer (CINOGY CMOS-1201, Germany), which can be equipped at the optical testing unit as shown Figure 1. For the gases at different temperatures, the light passes through the glasses in the aerosol tube and images on the CCD of CMOS-1201 to detect the spot of the light beam in real time.

After eliminating the background noise, the spot image at different temperatures can be acquired. When the aerosol tube is made of 304 stainless steel, the measurement results are shown in Figure 6. It can be found that the shape of the light spots is distorted and shows the imperfection of the imaging with an increase in gas temperature. That's due to thermal expansion along the airflow direction caused by the change of gas temperature, which results in the shift of light apertures position to obscure the partial light beam. A bad result must be acquired for smaller particle size and lower particle count efficiency because of the decreased incident light. Meanwhile, the shape of the OMV was changed to greatly reduce the performance of OPC detection by a failure of eliminating the error of edge detection in optical system.

With the aim to eliminate the above mentioned disadvantage, the invar material with a linear expansion coefficient of 1.6×10^{-6} m/°C was adopted for high-temperature experiment. The results are shown in Figure 7. At different temperatures, no deformation was found in the shape of light spots, indicating that the material can meet the testing requirement for high temperature up to 650 °C with a good heat resistance.

Table 1. Comparison of the application conditions of the two materials

Material	Light divergence angle /°	Gas temperature /°C
304 (Stainless steel)	<10	<200
4J36 (Invar)	<30	<700

The application conditions of the two materials are

summarized in Table 1. Compared with the stainless steel, the invar material is of great significance for the development of high-temperature detection device.

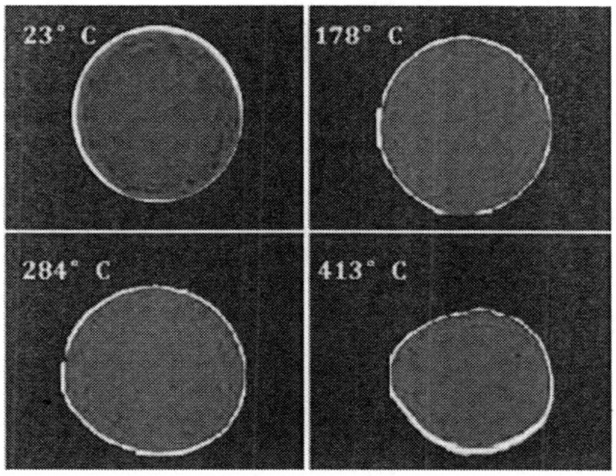

Figure 6. The spot images measured at four different temperatures of gas for 304 stainless steel.

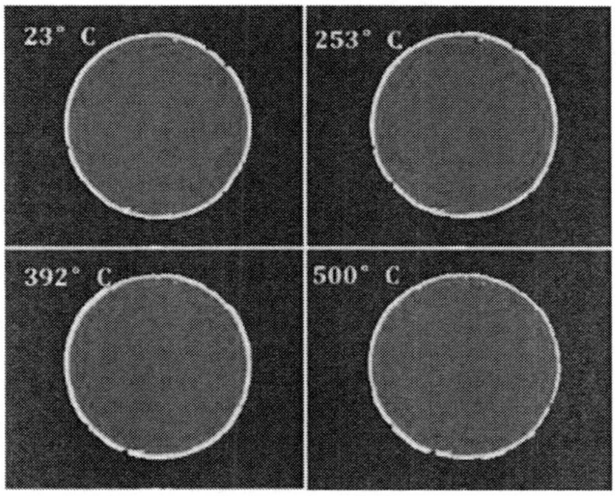

Figure 7. The spot images measured at four different temperature of gas for 4J36 (Invar).

3.2 Influence of temperature on optical system

In general the optical sensing system in the particle detection device is only applied at atmospheric temperature, but can also satisfy the temperature up to 150 °C. Under high-temperature condition, the gas volume is going to expand with the reduction in the gas density. The transmission characteristic of light beam could be changed, resulting from the decrease in the optical performance of the sensor. In order to investigate the influence of temperature on optical system, the specific causes are analyzed by theoretical method, and the extent of its influence is determined through experimental verification.

3.2.1 Theoretical method

By calculating the refractive index of the gas at different temperatures, the change of the OMV with the temperature can be obtained.

a) Calculation of gas parameters

The gas medium is air, so its refractive index can be calculated by the formula proposed by Rueger at the working temperature, given by:

$$n_{ph} = \frac{273.15}{1013.25} \cdot P \cdot N_{sph} \cdot \left(a_0 + \frac{a_1}{T} + \frac{a_2}{T^2} + \frac{a_3}{T^3} + \frac{a_4}{T^4}\right) \times 10^{-6} + 1 \quad (1)$$

Where P is the gas pressure; a_0, a_1, a_2, a_3 and a_4 are the correlation coefficient and constant; N_{sph} is the gas refractive index at atmospheric temperature, given by:

$$N_{sph} = (n_{sph} - 1) \times 10^6 = 287.6155 + \frac{1.62887}{\lambda^2} + \frac{0.01360}{\lambda^4} \quad (2)$$

The wavelength range is 400-700 nm. The gas refractive indexes at different temperatures in the range of 25-650 °C are calculated by Eq. 1 and 2. The results are shown in Figure 8. It can be found that the refractive index of air decreases with the increase in the gas temperature. And it can be inferred that the light divergence angle at high temperature is going to increase compared with that at atmospheric temperature. Consequently, the position of the beam focus is offset, leading to larger optical measuring volume, as shown in Figure 9.

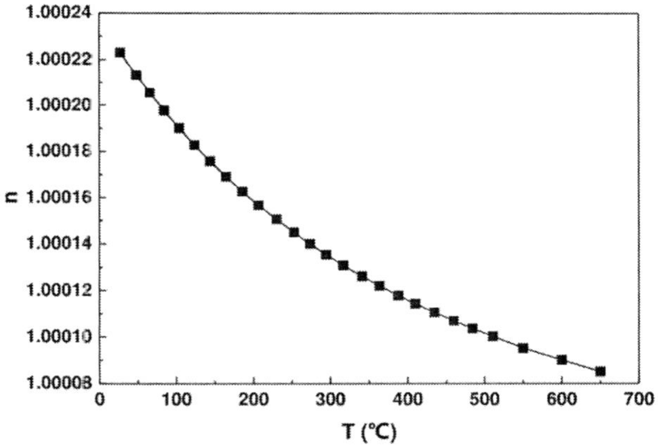

Figure 8. Gas refractive index at different temperature

In order to analyze the above unfavorable influence, a model of the OMV and gas temperature is needed to be built, mainly including the functions of the determination of the optical parameters and the calculation of the dynamic variable.

b) Calculation of optical parameters

On the basis of the Lorentz-Lorenz formula and actual gas state equation, the relationship between gas pressure and refractive index is given by:

$$\frac{n^2-1}{n_0^2-1}=\frac{\rho}{\rho_0}=\frac{PT_0Z_0}{P_0TZ} \tag{3}$$

Where T_0, P_0, ρ_0, Z_0 and n_0 are the gas temperature, pressure, density, compression factor and refractive index under atmospheric condition, respectively; T, P, ρ, Z and n are the gas temperature, pressure, density, compression factor and refractive index under high-temperature condition, respectively.

The light divergence angle is calculated by the refraction law shown below:

$$n_0^{'}\sin\alpha = n\sin\beta \tag{4}$$

Where n_0' and α are the air refractive index and light divergence angle under atmospheric condition; β is the light divergence angle under high-temperature condition.

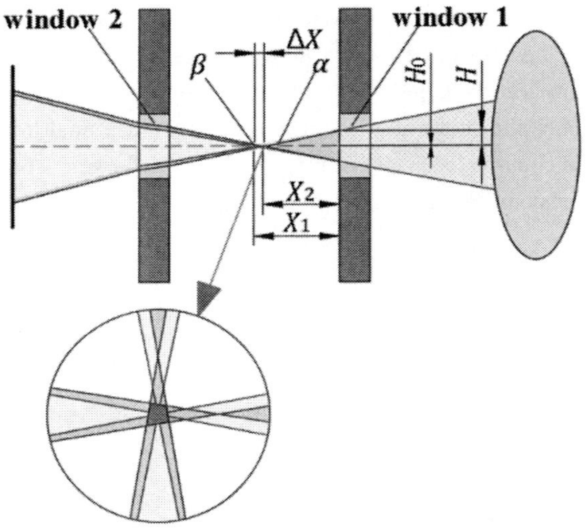

Figure 9. Schematic of the enlargement of the OMV

c) Calculation of dynamic variable

Based on the geometric optical theory and the mathematic relationship, the distance between the light focus position and the glass window 1 is triangular with the equivalent spot radius of the inner side of the glass window 1. According to X_1 and X_2 (the distance between the light focus position and the inner side of the glass window 1 under atmospheric condition and high-temperature condition), the theoretical model of the OMV with gas temperature was built. X_1 and X_2 are calculated by:

$$X1=\frac{H-H_0}{\tan\alpha} \tag{5}$$

$$X2=\frac{H-H_0}{\tan\beta} \tag{6}$$

The change of the light focus position (ΔX=$X1$-$X2$) can be calculated by Eq. 5 and Eq. 6.

The change of the equivalent spot diameter of the OMV can be calculated based on the triangle relation, as shown below:

$$\Delta D = 2\Delta X\tan\beta \tag{7}$$

Eq. 6 and Eq. 7 are the formulas of the theoretical dynamic model of the OMV with the change of the gas pressure.

Using the model, the optical parameters at different gas temperatures in the range of 25 ~ 700°C were calculated with three light divergence angles (4.02°, 9° and 13°, tested under atmospheric condition, respectively). The results are presented in Figure 10. It can be found in Figure 10 that ΔX increases with the gas temperature with approximate two patterns. In addition, ΔD changes by a smaller margin (less than 1 um). However, ΔD and ΔX satisfy a trigonometric relation. It can be also noted that the enlargement rate of the OMV volume (δ) has a small increase with the increase in gas temperature. Under the constant temperature, δ is going to increase for the larger light divergence angle, but it is still relatively smaller (<1%) compared with the OMV at atmospheric condition. Consequently, the influence of the gas temperature on the OMV can be ignored based on the above theoretical analysis.

3.2.2 Experimental method

In order to verify the reliability of theoretical analysis, an experimental dynamic model of the OMV with gas temperature variation was established based on the image analysis and the mathematical geometry relations, including the measurement of the change of light spot, the calculation of the geometric relation and the calculation of the relative variable.

By the refraction law, the changes of light spot on the inner and outer side of the glass window 2 under high-temperature condition (ΔD and $\Delta D'$) meet ΔD=$\Delta D'$. Taking the light focus position as the central axis, a triangle relation was formed by the distances between the light -

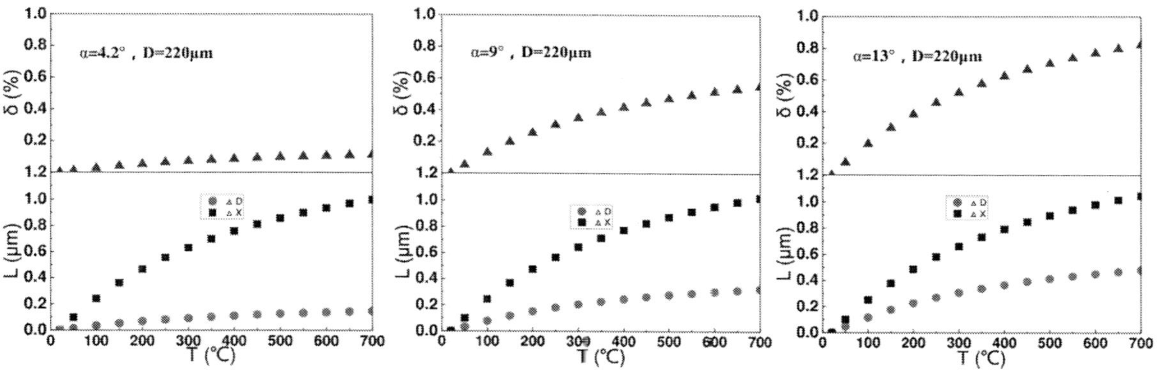

Figure 10. Results calculated based on theoretical dynamic mode with three light divergence angles (4.02°, 9° and 13°)

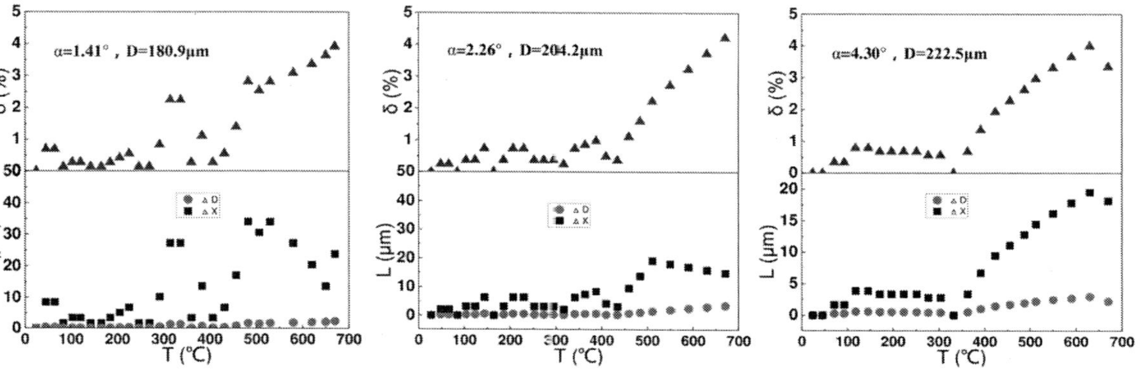

Figure 11. Results calculated based on experimental dynamic model

focus position and the inner side of the glass windows 1 and 2 (x and y), the difference of the light spot equivalent radius on the inner side of the glass windows 1 and 2 (H and h, $h = H - \frac{\Delta D}{2}$) and the equivalent radius of the OMV (H_0). Therefore, we have the following equations.

$$\frac{H - H_0}{h - H_0} = \frac{y}{x} \tag{8}$$

$$x + y = k \tag{9}$$

$$\beta = \arctan\left(\frac{h - H_0}{x}\right) = \arctan\left(\frac{H - H_0}{y}\right) \tag{10}$$

where β is the light divergence angle under high-temperature condition; k is the distance of the inner sides of the glass windows 1 and 2.

The changes of ΔX and ΔD under high-temperature condition to those under atmospheric condition are shown as follows:

$$\Delta X = \frac{y - x}{2}, \text{ or } \Delta X = y - \frac{k}{2}, \text{ or } \Delta X = \frac{k}{2} - x \tag{11}$$

$$\Delta D = 2\Delta X \tan\beta = \Delta D' \tag{12}$$

Eq. 11 to Eq. 12 are the formulas of the experimental dynamic model.

Using the above established experimental dynamic model, the ΔX, ΔD and δ of four cases in Table 2 were calculated, and the results are shown in Figure 11. It can be found that these three parameters increase obviously with the gas temperature increasing, but cannot meet the two function growth mode as the results of theoretical model. That may be caused by the reasons listed below: 1) the fluctuation of gas flow in the pipeline causes the unstable gas temperature field; 2) the beam quality analyzer is slightly influenced by other vibration sources; 3) the aerosol tube is affected by the gas flow fluctuation. But according to the experimental results, the maximum values of ΔX and ΔD are very small (less than 50 μm and 5 μm at the temperature below 700 °C, respectively) compared with the equivalent diameter of the OMV. Consequently, the ratio of the volume of the OMV is less than 5%. To sum up, the gas temperature has greatly weakened influence on the OMV, which can be ignored in actual technical development.

Table 2. Optical parameters of experimental model

number	light divergence angle /°	spot equivalent diameter /μm
1	1.41	180.9
3	1.94	285.1
4	2.26	204.2
5	4.30	222.5

3.2.3 Experiments

The performance of the optical sensor is investigated at different temperatures up to 650°C. The experimental results showed that the temperature of the optical sensor is around 50°C with the gas temperature in the range of 200-300 °C, and is going to increase about 10 °C at the gas temperature of 400-500 °C. The temperature is in the controllable range with a maximum value of 70 °C under the gas temperature of 650 °C. Therefore, the developed optical sensor can be used for the detection of particles in high-temperature conditions for a long time.

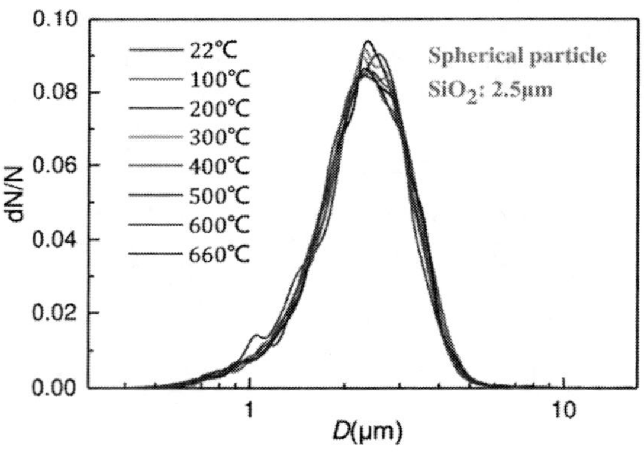

Figure 12. The results obtained under high-temperature condition

In the experiment of particle detection, the standard silica spherical particles with a strong temperature resistance ($D_{particle}$=2.5 μm) were tested at different gas temperatures. The testing time is 60 s. Figure 12 shows the test results. It can be found that all of the particle size distributions are in good agreement and have a good normal distribution, which meets the test results of single particle size, and the peak particle size is about 2.5 μm. Consequently, the optical sensor worked well for the particle detection under higher-temperature condition.

4 Conclusion

The emission of high-temperature flue gas causes serious air pollution and partial waste of available heat. In order to meet the relevant emission standards, additional purification units (high-temperature ceramic filters, etc.) are installed in the industrial process to collect the dusts in the flue gas. As the purification units are important for controlling the dust concentration, it is necessary to calculate the separation efficiency through a real-time measurement of the dust concentration in the flue gas in the upstream and downstream of the purification equipment.

In present work, a single particle measurement method and device suitable for high-temperature condition were developed. In order to complete the particles measurement at 650 °C, the beam images in optical sensing system under different temperature conditions were measured in real time by CMOS sensor. It was found that the beam images were deformed when the light divergence angle increased to more than 10°, and the deformation became more serious as the temperature increased. Depending on the principle of heat transfer, thermal expansion along the airflow direction caused by heat conduction between the duct and the gas is the main cause of this phenomenon. By changing the aerosol tube material t o one with an ultra-small linear expansion coefficient (Invar: 4J36), the optical system can be applied to the condition with temperature up to 700 °C, and in particular the light divergence angle extends to 30°.

According to the principle of gas state equation, the optical parameters (refractive index) of the gas at different temperatures change with the change in its state parameters. According to the results calculated by the formula of Ruegur's atmospheric refractive index, it was found that the refractive index of the gas decreases with the increase in temperature, which leads to the increase in the divergence angle of the beam in the aerosol tube, and shift of the position of the optical beam. Based on the geometrical optics theory and the geometric mathematical relationships, the theoretical dynamic model of the optical measuring volume with the change of the gas temperature was established. By calculating the amount of the beam focus offset and the volume of the optical measuring volume at different temperatures, it was found that the optical measuring volume increases with the temperature increasing. However, when the gas temperature was 650 °C, the increase rate of optical measuring volume was extremely small, so the influence of the temperature on the optical system can be negligible.

The performance of the whole detection system was tested by using 2.5 μm standard spherical silica particles. The satisfying measuring results were obtained and they could be applied to the actual on-site measurement at the temperature of 650 °C.

Acknowledgement

The authors thank the supports of "The National Key Research and Development Program of China (NO.2016YFB0601100)"; and would also like to thank the laboratory and staff of the Beijing Key Laboratory of Process Fluid Filtration and Separation, China University of Petroleum.

Reference

[1] J. Meyer, H. Umhauer, K. Reuter, A. Schiel, G. Kasper, M. Förster, Concentration and size measurements of fly ash particles from the clean gas side of a pressurised pulverised coal combustion test facility, Powder Technology, 180 (2008) 57-63.

[2] H. Leibold, F. Dirks, V. Rüdinger, Particulate emissions from a LLW incinerator and off-gas cleaning with a new type of ceramic candle filter, Waste Management, 9 (1989) 87-94.

[3] B.L. Perkins, Incineration facilities for treatment of radioactive wastes: a review, Incinerators, (1976).

[4] D. Fino, N. Russo, A. G. Saracco, V. Specchia, Multifunctional Filter for Treatment of the Flue Gases from Municipal Waste Incinerators, Industrial & Engineering Chemistry Research, 44 (2005) págs. 9542-9548.

[5] Scheibner, Wolters, Schumacher hot gas filter long-term operating experience in the nuon power buggenum IGCC power plant, (2002).

[6] J.M. Wang, Shell coal gasification technology and its application in China, Coal Quality Technology, (2006).

[7] Q. Xu, Z. Ji, L. Yang, Performance assessment of high temperature flue gas on-line particle size analyzer, Ciesc Journal, 63 (2012) 3506-3512.

[8] B.Y.H. LIU, V.A. MARPLE, K.T. WHITBY, N.J. BARSIC, Size distribution measurement of airborne coal dust by optical particle counters, Aihaj, 35 (1974) 443-451.

[9] R. Maus, F. Löffler, Size analysis of airborne bacteria by an optical particle counter and an impactor, Journal of Aerosol Science, 25 (1994) 141-142.

[10] B. Sachweh, H. Büttner, F. Ebert, Improvement of the resolution and the counting accuracy of an optical particle counter by fast digital signal recording, Journal of Aerosol Science, 20 (1989) 1541-1544.

[11] E.W. Max Born, Principles of Optics, Cambridge University Press, Cambridge, 1999.

[12] S. Hochrainer, Further developments of the particle counter sizer PCS-2000, Journal of Aerosol Science, 31 (2000) 771-772.

[13] B.S. Li, Z.L. Ji, L. Mei, An approach for correcting particle size distributions measured by the optical particle counter welas 2000, Aip Conference, 2010, pp. 686-689.

[14] J. Stier, M. Quinten, Simple refractive index correction for the optical particle counter PCS 2000 by Palas, Journal of Aerosol Science, 29 (1998) 223-225.

[15] H. Umhauer, Particle Size Distribution Analysis by Scattered Light Measurement Using an Optically Defined Measuring Volume, Journal of Aerosol Science, 14 (1983) 765-770.

[16] B. Sachweh, H. Umhauer, F. Ebert, H. Buttner, R. Friehmelt, In situ optical particle counter with improved coincidence error correction for number concentrations up to 107 particles cm-3 quantification and elimination of the effect of particle shape and structure, Journal of Aerosol Science, (1998) 1075-1086.

[17] D.I.J. Raasch, D.I.H. Umhauer, Errors in the Determination of Particle Size Distributions Caused by coincidences in optical particle counters, Particle & Particle Systems Characterization, 1 (1984) 53-58.

[18] H. Buettner, Measurement of the Size of Fine Nonspherical Particles with a Light-Scattering Particle Counter, Aerosol Science & Technology, 12 (1990) 413-421.

[19] D. Holve, S.A. Self, Optical particle sizing for in situ measurements Part 2, Applied Optics, 18 (1979) 1632-1645.

508245 An Experimental Investigation of Single Droplet Drying Above Boiling Point

Abstract:

An experimental method has been developed to enable dried particle structure, properties and drying rate to be predicted based on droplet drying history, particularly focusing on effects driven by boiling including bubble nucleation, growth and collapse. The method involves a single droplet drying rig which will be used to investigate drying kinetics above boiling point and across a broad range of drying conditions, drop size and composition space. The drying rig was improved as compared to previous apparatuses reported in literature, in that it is designed to measure droplet drying rates and morphological changes at high temperatures and under conditions that more closely resemble those found in industrial applications. The aim is to conduct drying experiment for droplets as small as 300μm diameter, in an air stream which temperature may be as high as 350°C.

Drying kinetics and morphological evolution will be determined, particularly focusing on the second drying stage as well as inflation/deflation periods. Initially, the constructed drying rig is checked and evaluated by conducting experiments on droplets of pure water. Afterwards, experiments will be conducted on solutions of different materials such as Hydroxypropyl Methylcellulose (HPMC), sodium sulphate and sucrose. An assessment of morphologies, transformations and drying behaviour under isothermal drying would be completed at a wide space of process condition, initial water content, initial drop size and composition space. The investigation will look to link material properties, particle size, initial moisture content and drying history to morphology. Results will be compared with single droplet drying models reported in literature.

An initial investigation has focused on mapping the morphological drying behaviour of HPMC, sodium sulphate and sucrose. It has shown that the three materials show different behaviour; sucrose showed deflation and inflation cycles, sodium sulphate droplets supersaturated and crystallized while HPMC showed inflation and rupture. Differences in drying behaviour is due to difference in material and transport properties. Therefore, the next stage of this work is focused on conducting some physical properties measurement at different water concentrations and temperature. In addition, different materials will be dried on the rig and the drying behaviour will be determined. The experimental data from this work will be used to develop, a universal morphology regime map which will link material properties, particle size, initial moisture content and drying history to morphology.

508482 Characterising Powder Flow in Dynamic Processes

Abstract:

Flow inconsistency of cohesive powders is a major issue for process industries. It is often difficult to predict and control flow of cohesive powders to enable efficient and reliable process operations particularly for dynamic processes such as mixing and filling. Many quantitative and qualitative flow measurement techniques exist that are used for measuring powder flowability under quasi-static conditions. However, there are few available techniques capable of assessing powder flowability under dynamic conditions. The FT4 Powder Rheometer (Freeman Technology), which drives a rotating blade along the vertical axis through a column of powder contained within a cylindrical vessel, is one of the few available devices that is capable of measuring powder flow in a dynamic state [1]. The flow resistance of the rotating blade as it penetrates the powder bed is characterised by flow energy, however, there is lack of understanding in how to relate this measurement to process design.

This work focuses on determining the relationship between the torque and shear stress in the FT4. The effect of particle size on the external flow profile around the blade has been investigated experimentally for glass beads using a high speed camera. The material investigated shows a decrease in downward specific flow energy with increasing particle size; indicating that the increased shear resistance in the bulk for smaller particle sizes exists under dynamic conditions. Particle Image Velocimetry (PIV) analysis indicates an increase in the volume of particles displaced by the blade with increasing particle size for both upward and downward blade movement. An increase in impeller tip speed leads to a reduction in flow energy for this material, and a reduction in the volume displaced by the blade. The prevailing stresses and strains as well as the volume of particles displaced by the blade are estimated by Discrete Element Method (DEM). DEM allows for the determination of a relationship between the flow energy and/or shear stress with particle properties and blade velocity and direction. Furthermore, Positron Emission Particle Tracking (PEPT) analysis of the FT4 has been carried out to determine the radial and axial displacement of particles with respect to the bed height, and to validate the DEM results.

508671 Impact of Non-Spherical Projectiles on Granular Media

Abstract:

Recently, several studies involving numerical and experimental methods have focused on the study of impact dynamics in both dry and wet granular media. An example of such instances can be an animalâ€™s foot impacting sand during its locomotion. Understanding the physics behind the interaction between granular media and animal feet can be helpful in optimizing locomotion parameters for bio-inspired robots. However, most of the studies involving impact dynamics considered spherical projectiles under different conditions, while characteristic models should involve more complex shapes. Impacting different geometries with conserved density, volume, and velocity on a granular bed may experience contrasting drag forces upon penetration. This is due to the difference in the surface areas coming in contact with the granular media. In this work, non-spherical geometries of following shapes were considered: cylinder, triangular prism, hexagonal prism and cuboid. These geometries were impacted on a loosely packed non-cohesive dry granular bed with the same impact velocities. The effect of surface area in contact with the granular media was analyzed for impact, with the help of penetration depth measurements. We also tested whether the power law for spherical projectiles is applicable for non-spherical projectile impacts.

508664 Computational and Experimental Shear Cell Study with Rigid Cylindrical Particles

Abstract:

Flow properties of particles are commonly assessed through shear cells. Both the Schulze ring shear tester (SRST) and Jenike shear tester (JST) measure the interparticle stresses which are important to the design of many granular flow processes. This work uses precision-cut rigid cylinders of various aspect ratios and sizes to compare the measured stresses between both testers. It is found that the shearing results for both testers are similar, which is important for future work involving flexible biomass materials.

Previous work has sought to use shear cell results to validate various aspects of discrete element method (DEM) simulations. The SRST is recreated in a DEM simulation with rigid glued-sphere cylindrical particles to assess the contact detection method by comparing the experimental and computation results. Also, a parameterization study is performed to examine the effects of coefficient of restitution, coefficient of friction, and other particle properties on the measured steady-state stress. Overall, this serves as an efficient validation of the contact detection method which will be used to simulate complex particle shapes.

508865 Defluidization Behaviour of Industrial Reactive Powders at High Temperature

Abstract:

Fluidized bed reactors are used in a wide range of industrial chemical processes, and they are particularly useful in high temperature systems thanks to their ability to provide high heat transfer rates and rapid solids mixing which lead to isothermal and more controllable conditions. Their applications spread the environmental, chemical, energy and process industries. For example, they are a key technology in the petroleum industry, catalytic chemicals synthesis processes, combustion and gasification of solid fuels (coal, wastes and biomass), metals productions and many more.

Use of fluidized beds in combustion-type operations may lead to particles sintering and bed agglomeration, which again may result in defluidization, unscheduled process downtime and additional costs. However, the details of the physical mechanisms behind such phenomena are still not entirely understood. Despite the countless researches that have been done so far, such processes still disclose unrevealed features and challenge the researchers worldwide.

The objective of this paper is to assess the effect of high temperature on the flow properties and fluidization behaviour of mixtures of carbon coke and two different titanium ores, namely synthetic and natural rutile. To this end, a simple approach for the early detection of agglomeration in the regimes of incipient fluidization -where the carbon combustion causes defluidization- will be followed. Such a monitoring approach is based on the simultaneous measurements of local temperatures and the pressure drop across the particles bed.

507739 Manufacturing and Characterization of Spherical Blend PBT-PC Particles for Additive Manufacturing

Abstract:

Polymer powder-based additive manufacturing technologies often yield excellent mechanical and functional part qualities. One of these technologies is selective laser sintering (SLS), a process employing micron-sized polymer particles, which are selectively fused by a laser [1]. While hardly any boundaries regarding part design exist, there are quite some restrictions concerning the variety of commercially available SLS materials. For the time being, the most widely used polymeric material for SLS is polyamide 12 (PA12) with a market share of roughly 95% [2]. Hardly any blend materials are available; mostly , the processing of multi-material systems via SLS is achieved by simply mixing different polymer powder materials [3].

In this contribution, a new approach to blend particles is introduced. Via co-grinding and subsequent rounding, manufacturing of spherical particles consisting of semi-crystalline polybutylene terephthalate (PBT) and amorphous polycarbonate (PC) thermoplastic is possible. The polymers are jointly comminuted and mixed in a planetary ball mill and the product particles are rounded in a heated downer reactor [4,5]. The thereby achieved intermixed, but not transesterified, PBT-PC composite particles are characterized regarding size and morphology via electron microscopy. Intermixing of the polymers is investigated via staining techniques and Raman spectroscopy. Thermal properties of the blend particles are analyzed via DSC, also in regard to transesterification in SLS process-like conditions. Furthermore, flowability and size distribution of the blend particles are examined. This way, we show the high potential of our process route [4] for the production of novel polymer blend particle systems for SLS.

References:

[1] I. Gibson, D. Rosen, B. Stucker, Additive manufacturing technologies, 2. ed., Springer, New York, NY, 2015.

[2] T.T. Wohlers, Wohlers Report 2016, Wohlers Associates, Fort Collins, Col., 2016.

[3] D. Drummer, K. Wudy, F. Kühnlein, M. Drexler, Polymer Blends for Selective Laser Sintering: Material and Process Requirements, Phys. Procedia. 39 (2012) 509–517. doi:10.1016/j.phpro.2012.10.067.

[4] J. Schmidt, M. Sachs, C. Blümel, B. Winzer, F. Toni, K.-E. Wirth, W. Peukert, A novel process route for the production of spherical LBM polymer powders with small size and good flowability, Powder Technol. 261 (2014) 78–86. doi:10.1016/j.powtec.2014.04.003.

[5] J. Schmidt, M. Sachs, S. Fanselow, M. Zhao, S. Romeis, D. Drummer, K.E. Wirth, W. Peukert, Optimized polybutylene terephthalate powders for selective laser beam melting, Chem. Eng. Sci. 156 (2016) 1–10. doi:10.1016/j.ces.2016.09.009.

506658 Creating Tuneable Agglomerates via 3D Printing

Abstract:

It is essential to offer a universally accepted way to systematically evaluate agglomerate breakage propensity and mechanisms. Discrete Element Method (DEM) simulation is a commonly used tool but is limited by the lack of identical, controlled agglomerates to test and validate simple models, let alone replicate the complex structure of real industrial agglomerates.

Our research presents a systematic approach to produce 3D printed agglomerates with tuneable properties. Agglomerates with different mechanical properties and structures were designed and printed by an Objet500 Connex 3D printer. The tensile and shear strength of inter-particle bonds was characterized using Instron tester. Breakage results of agglomerates with simple cubic structures and the design of agglomerates with irregular structures were presented. This 3D printing method will allow more rigorous testing of agglomerate breakage models. Discrete Element Method (DEM) has been used to simulate the agglomerate crushing using the Timoshenko Beam Bond Model (TBBM) with bond properties matching the 3D printed agglomerates. For both two agglomerate structures used in this research, the simulation and experiment showed similar breakage patterns. However, the simulation results significantly underestimated the compressive loads of spherical random structured agglomerates. In order to improve the simulations of compressive granule breakage, the influences of bond geometry definitions and non-linear behaviour of the bond contacts need to be explored.

508699 Effect of Particle Size Distribution and Wettability on Penetration Behavior, Granule Formation and Granule Properties in Single Drop Granulation

Abstract:

In wet granulation, single drop granulation (where one drop forms one granule) can simplify the complexity and reveal the fundamentals of the interaction between the liquid and the powder bed. Combining drop penetration time, granule formation mechanisms and the characterization of granule properties is a novel study to achieve a comprehensive map of the single drop granulation process, which is relevant in particulate applications like pharmaceuticals, detergents, agricultural chemicals and catalysts.

In this work, single drop impact of liquid on a static powder bed was studied to investigate the granule formation mechanism, droplet penetration time, and the characterization of granule properties (morphology, internal structure and content uniformity). Water was used as the liquid binder and two pharmaceutical powders, microcrystalline cellulose (MCC) (mean particle size ~54.8 Î¼m, contact angle with water ~36Â°) and acetaminophen (APAP) (mean particle size ~9.74 Î¼m, contact angle with water ~89°), were mixed to make heterogeneous powder beds with varying compositions. The complete drop impact and penetration was recorded with a high-speed camera. The granule morphology, its internal structure and content uniformity were characterized by the prism method with image analysis, micro-CT and UV-vis spectrometry, respectively. Besides the current MCC and APAP system, MCC and APAP with other particle sizes and relative contact angles will also be studied. It is believed that the mean particle size of the powder bed is predominant in influencing the granule formation mechanism, drop penetration time, and granule properties.

508017 Atomically Deposited Sintering Aids: Assessing the Effects of Alumina Particle ALD on the Sintering and Performance of SOFC Electrolytes and Dental Ceramics

Abstract:

The crystal structure of zirconium oxide (ZrO_2) can be tailored by the incorporation of aliovalent dopants, most commonly, yttrium oxide (Y_2O_3). The concentration of dopant dictates the thermodynamically stable crystal phase where 3 mol%
Y
conducting cubic phase. To achieve suitable mechanical properties and ionic conductivity, Y-TZP and YSZ ceramics must be near theoretical density, requiring high sintering temperatures. Previous researchers have successfully lowered the required Y-TZP and YSZ sintering temperatures through incorporation of aluminum oxide (Al_2O_3) particles. While this technique has proven to reduce the sintering temperature in both systems, the presence of particulate Al_2O_3 inclusions in densified parts reduces the homogeneity of the material which may have deleterious effects on densification and microstructure. In this work, we report on the use of atomic layer deposition (ALD) to conformally coat individual Y-TZP and YSZ particles with the desired amount of Al_2O_3. Constant-rate-of-heating (CRH) experiments were conducted using a horizontal push-rod dilatometer to extract the activation energy of sintering and also gain mechanistic insights into the active diffusion mechanisms as a function of ALD film thickness. Transmission electron microscopy (TEM) and scanning electron microscopy (SEM) were used to characterize the as-deposited thin Al_2O_3 films and the effects of the Al_2O_3 films on grain growth and microstructure during densification. Ionic conductivity and flexural strength tests were conducted to determine the effect of Al_2O_3 thin films on relevant material properties. Al_2O_3 incorporation by ALD is shown to have unique phenomenological effects on the densification of Y-TZP and YSZ as compared with conventional Al_2O_3 incorporation using particle-based approaches.

508644 Effect of Particle Size Distribution and Operating Parameters on Conduction and Convection Heat Transfer Mechanisms in Rotary Drums

Abstract:

Granular materials undergo process steps that include transportation, drying, heating and chemical or physical conversion. In several instances, these materials need to be heated or cooled during processing, and rotary drums (kilns) are the most commonly used process equipment for this purpose. It is not always possible to conduct experiments using different materials and consider all process conditions to study the thermal mechanism due to the inherent complexity of these processes. Therefore, computational modeling and simulations have been extensively used to understand the heat transfer mechanism in rotary drums. To maintain simplicity and to overcome the challenges of simulating polydispersed particles, a common practice is to assume that the polydispersed particles with a narrow size distribution behave like monodispersed particles thermally. However, no effort has been made to validate this assumption. This work focuses on understating the effect of polydispersed particles on the conduction and convection heat transfer mechanisms in indirectly heated rotary drums. It also aims to investigate the effect of rotation speed, fill level, particle size and their interaction effects on the conduction and convection heat transfer mechanisms in rotary drums. For this, the discrete element method (DEM) (using MFIX-DEM, an open-source computational fluid dynamics solver suite) simulation technique is used to analyze the thermal behavior of spherical silica beads, which are the most common catalyst support material.

507974 Selective Hydrogenation of Citral over Supported Pt Catalysts on Various Substrates

Abstract:

Highly dispersed platinum (Pt) nanoparticles (NPs) were deposited on various substrates by atomic layer deposition (ALD) for the reaction of citral hydrogenation. The substrates included multi-walled carbon nanotubes (MWCNTs), silica gel (SiO_2), commercial Î^{3}-Al_2O_3, and ALD-prepared porous Al_2O_3 particles (ALD-Al_2O_3). Based on TEM and CO-chemisorption analysis, the Pt NPs on all substrates were around 1.3 nm, and Pt/MWCNTs, Pt/ALD-Al_2O_3 and Pt/Î^{3}-Al_2O_3 showed similar Pt dispersion (75%-85%). But the Pt dispersion on SiO_2 was low (63%). In the reaction of citral hydrogenation, Pt/SiO_2 showed the highest activity due to the strong acidity of SiO_2 and the conversion of citral reached 82% after 12 hr with a selectivity of 58% of unsaturated alcohols (UA), including geraniol and nerol. Among all catalysts, Pt/MWCNTs showed the highest selectivity of UA (65%) due to its unique structure and electronic effect. In the cycling test, Pt/MWCNTs and Pt/ALD-Al_2O_3 catalysts were more stable than Pt/SiO_2 which could be attributed to the different interactions between the catalysts and the solvent.

508345 Numerical Simulation of Particle Classification in a Classifier Based on Coanda Effect

Abstract:

Efficient and precise classification of particles has been an important issue in many industrial applications as well as in academia. A relatively new particle classifier based on 'Coanda effect' was developed and reported to have many favorable features (Liang 2010). They include simultaneous multiple classification, simple design with no moving parts, and easy scale up to higher production rates. However, this classifier has attracted little attention so far, and our understanding of underlying mechanism associated with classification performance needs to be improved.

The purpose of this study is to investigate the detailed fluid flow and particle transport characteristics in a Coanda classifier using CFD (Computational Fluid Dynamics). The main focus is placed on the effects of working conditions (inlet flow velocity, particle diameter) and numerical method for turbulence dispersion of particles on the classification performance. As a numerical method, three different approaches are considered: k-ε model, k-ε model combined with DRW (Discrete Random Walk) model, LES (Large Eddy Simulation). In the first approach, standard k-ε model is adopted to obtain the mean flow and then particle trajectory is calculated based on the mean flow without consideration of turbulence dispersion. However, in the second approach using DRW model, the instantaneous flow velocity is constructed under the assumptions of isotropic turbulence in order to include turbulence effect. Finally, in LES, the instantaneous flow field is dynamically and accurately obtained at each time step. Numerical simulations are carried out for two-dimensional mean flow in case of k-ε model, but for three-dimensional unsteady flow in LES. Particles are tracked using a Lagrangian approach based on one-way coupling. We consider talc particles with density of 2,860 kg/m^3 and diameter of 1 to 20 μm.

The Coanda classifier is mainly composed of three inlet channels, three outlet channels, and Coanda block. Inlet channels are divided into one main inlet and two control inlets. The main air jet containing particles is supplied with the velocity of 35 m/s through the main inlet. Two control air jets with velocities of 32 m/s (lower) and 40 m/s (upper) are also supplied through the control inlets for enhanced control.

The trajectories of moving particles obtained from the mean flow (k-ε model) are analyzed for different diameters. The main jet containing particles deflects toward the curved Coanda block due to Coanda effect as expected. Fine particles (d≤2μm) follow the flow streamlines and enter the lower outlet channel. However, coarse particles (d≥8μm) are detached from the streamlines and go into the upper outlet channel. This observation agrees well with the previous result (Liang, 2010).

Even though the mean flow and corresponding particle trajectories provide useful information, realistic particle passage do not take a form of single curve but they change in time due to flow turbulence. Therefore, particle trajectories are simulated again, turbulence dispersion being considered with DRW model. Due to turbulence dispersion, particles of the same diameter are scattered around the particle pathline obtained from the mean flow. As a result, particles of the same diameter can not be collected into a single outlet. For example, some particles (about 5%) of 1 μm enter the middle outlet even though most of them (about 95%) enter the lower outlet designed for collecting fine particles. This means that flow turbulence deteriorates the accuracy of classification.

Finally, LES simulations are performed to understand dynamic behaviors of fluid flow and particle transport. Instantaneous flow fields such as vorticity contours clearly show unsteady fluid motions inside the classifier. It is also shown that the generation of unsteady vortices leads to the scattering of particle trajectory.

A more detailed analysis including the classification efficiency will be given in the final presentation.

Acknowledgements

This research was supported by the National Research Foundation of Korea (NRF) under grant number 2017M2A8A4018482 and 2015R1D1A1A0105 9675.

References

Liang, S. (2010). *Chem. Eng. Comm.*, **197**, 1016-1032.

503491 Numerical Simulation of the Secondary Air Distribution Layer in the Gas Solid Fluidized Bed Based on Flunt Software

Abstract:

Due to the continuity beneficiation in the practical production, the gangue would accumulate on the distributor forming the gangue layer, which was called Secondary Air Distribution Layer and has vital effect on the gas distribution. In this paper, the ventilation characteristics of the secondary air distribution layer was analyzed. In addition, FLUENT software was used to study the effect of the modeled layer for the fluidity characteristics (e.g., pressure drop, bed height, volatility, etc.) in the gas-solid fluidized bed with the complex layer. The results showed that the complex layer could promote the secondary distribution of the airflow in the bed, which effectively inhibited the production of large bubbles and reduced the bed pressure drop fluctuation. Thus, itâ€™s of significance to provide synergy effect on the fluidization stability in the fluidized bed. Based on the above analysis, the introduced complex structure would be convenient to achieve the coal separation effectively.

507949 CFD-DEM Simulation of the Fluidization of Non-Spherical Particles in Fluidized Bed

Abstract:

In chemical and energy industries, fluidization concerning non-spherical particles (such as rod-like particle, disk-like particles, ellipsoidal particles, etc.) is often encountered, but research in this area is minimal. Yet understanding fluidization of non-spherical particles is crucial in many relevant industrial process. To investigate the fluidization of non-spherical particles, a CFD-DEM coupling algorithm is employed in this research to simulate a bubbling fluidized bed. The non-spherical particles are modeled by super-ellipsoids, multi-sphere models, and other particle models. The simulations results show that, the fluidization of non-spherical particles has significant differences from that of the spherical particles. The local void fraction and particle's projected face area perpendicular to fluid flow direction are directly related to the orientations of the non-spherical particles, which has a significant influence on the particle drag force and fluidization. Due to the significance of particle orientation, we further investigate the factors that influence the distribution of particle orientations. The results show that particle orientation depends on the fluidization time, fluidization velocity, and structure of the fluidized bed.

508320 CFD and DEM Simulation of the Cold Spray Process for Surface Coating with Fine Particles

Abstract:

The coating and functionalization of technical surfaces has been become increasingly important in the last two decades for various applications, since of its manifold possibilities for process optimization, as friction reduction, corrosion resistance or improved wetting behavior. Cold spray is a thermal spraying method for surface coating. It was developed in the late 1980s in Novosibirsk [1] and has gone through continuous optimization on certain applications since then. For example showed our previous study a reduction of the friction properties on steel pins, microstructured with fine Titania particles by an adapted spraying method [2]. In the cold spray process particles are dispersed in a gas stream, which is heated up to several hundred degrees and accelerated to supersonic velocities in a Laval nozzle. The particles impact on a substrate with about 800 m/s and form a strong bonding due to the high kinetic energy which is transferred to deformation and heat in the contact zone. In contrast to other thermal spraying methods, the temperatures occurring during the process are far below the melting temperatures of the used materials, which makes the process suitable also for thermal sensitive powders. The quality of the produced coating layers depends highly on the properties of the powder and substrate as well as on the nozzle design and the process parameters as initial temperature and particle velocities. Due to the high gradients, the high occurring velocities and the small size of the particles, typically in the range of 1 to 50 µm, the contact zone of impacting particles can only be measured indirectly and with destructive methods. Therefore, still not all phenomena which lead to a strong bonding in the contact area and a high quality coating are completely understood. For the study of these mechanisms two different simulation approaches have been used:

a) Simulation of particle and gas flow with Computational Fluid Dynamics (CFD)

For a better understanding of the spray process, the nozzle, the adjacent flow areas and the substrate were computed with a CFD method. The trajectories of single particles were calculated by an Euler-Lagrange approach. For the flow the k-Ï‰ SST model was used. Local temperatures in the substrate were calculated by the law of heat conduction. The impact conditions dependent on the nozzle geometry and the particle properties. With the local temperature distribution in the substrate and in the particle, precise impact conditions could be estimated. A study has been conducted to vary the parameters of the spray process, temperature and pressure as well as the particle properties as shape factor, density and size. Moreover the nozzle geometry has been varied in the simulation to optimize the spray process. The simulation results are compared to images of coated surfaces obtained by Scanning Electron Microscopy (SEM) for validation.

b) Simulation of particle-surface impact behavior with Discrete Element Method (DEM)

The DEM is widely used for investigations of particular processes. The advantage of this method is the detailed calculation of particle-particle or particle-wall contacts to obtain the important parameters as forces and deformations. The contact force in the particle wall contact is also dependent on the surface morphology. Therefore the surface topography of two real technical surfaces, a polished and an unpolished Titanium substrate were measured by Scanning Probe Microscopy (SPM) imaging with a nanoindenter (TI Premier, Hysitron). The surface topography was discretized and implemented in a DEM model. A parameter study has been conducted to determine the distribution of forces and contact angles occurring at the particle impact. Different particle sizes and impact velocities were studied.

The cold pray process dependents on various parameters which affect each other. Since of the high gradients and the small scales occurring during the process experimental measurements are often difficult. With this numerical study different parameters could be considered uncoupled of the other parameters. The results can be used for a better process understanding and an optimization of commercial cold spray systems.

[1] Alkhimov, A. P., Papyrin, A. N., Kosarev, V. F., Nesterovich, N. I., Shuspanov, M. M.: Gas dynamic spraying method for applying a coating; US Patent, 5302414; 1994

[2] Buhl, S., Schmidt K., Sappok D., Merz, R., Godard, C., Sauer, B., Kerscher, E., Kopnarski, M., Antonyuk, S., Ripperger, S.: Surface structuring of case hardened chain pins by cold-sprayed micro particles to modify friction and wear properties; Particuology 21, 2015

508568 Two-Fluid Validation of Constituive Models for the Simulation of Cylindrical Particles

Abstract:

Two-fluid simulations are a popular tool in the modeling of large processes. The constitutive models that close the governing equations are based in granular kinetic theory and typically assume that the particles are spherical. In reality, very few processes utilize spherical particles, hence there is a need for constitutive models that describe more complex shapes. Previous work used discrete element method (DEM) simulations to create a solid stress[1,2] and collisional dissipation rate[3,4] model for cylindrical particles that can replace the current spherical models. These new models have been added to MFiX and this work focuses on their validation. Cylindrical particles of various aspect ratios and solid densities were used in cratering and hopper discharge experiments. Simulations are designed to recreate the experiments in order to compare the results for crater depth and hopper discharge rate. The model effectiveness is assessed with regards to assumptions made concerning particle rotation and preferential alignment.

1. Guo Y, Wassgren C, Ketterhagen W, Hancock B, James B, Curtis J. A numerical study of granular shear flows of rod-like particles using the discrete element method. *J Fluid Mech*. 2012:1-26.

2. Berzi D, Thai-Quang N, Guo Y, Curtis J. Stresses and orientational order in shearing flows of granular liquid crystals. *Phys Rev E*. 2016;93(4):40901.

3. Buettner KE, Guo Y, Curtis JS. Using the Discrete Element Method to develop Collisional Dissipation Rate Models that Incorporate Particle Shape. *AIChE J*.

4. Berzi D, Thai-Quang N, Guo Y, Curtis J. Collisional dissipation rate in shearing flows of granular liquid crystals. *Phys Rev E*. 2017;95(5):50901.

509002 The Comparsion of Coarse Grained CFD-DEM for Simulating the Dense Bubbling Fluidized Bed" Requires Immediate Action

Abstract:

CFD-DEM is an effective tool to study the hydrodynamics in dense bubbling fluidized bed. However, the expansive cost of calculating the particles motion is not easy for its application in industrial application or dense particles system. Thus several coarse grained models (CGM) were put forward to simplifying the calculation of particles and decreasing the particles motion, such as PSM, SPA, MSPA,CGEE, EMMS-DPM where a group of particles are set as a parcel or coarse particle to reduce the computation. In this study the experiment and simulation test conducted to test five coarse grained methods. The result indicated that the models increasing the gravity force, pressure gradient force and drag force linearly with the coarse grained ratio (CGR) can be effective to ensure similar macro behavior between original system and coarse graining system. The simulation data of the solid fraction distribution and bed expansion ratio reflected good agreement as experiment. But the bubbles structure became less clear as increasing the CGR. Furthermore, the power spectrum density (PSD) and Wavelet analysis of the pressure signal were used and showed that SPA and EMMS-DPM gave better agreement result in term of dominant frequency and energy ratio at mesoscale feature. Finally the computational efficiency for different models was compared and the models of EMMS-DPM, PSM and CGEE have higher computation efficiency than SPA and MSPA relatively if the collision model of spring-damping were adopted.

509153 Stability Analysis of Uniform Gas Solids Flow

Abstract:

Uniform distribution of gas solids flow should be ensured to gain sufficient performance of fluidized beds. The highly chaotic entropy dissipation across solids bed will trigger non-uniform fluidization, although distributor plays an important role in obtaining homogenous gas stream. Hence, total solids pressure loading (Î¦$_T$), the pressure drop ratio between solids bed and distributor, is introduced here to present their interaction. The boundary between uniform and non-uniform fluidization can be detected by stability analysis based on Prigogineâ€™s minimum of entropy production principle, and this criterion for *N* parallel paths can be simplified into calculating the second derivative of pressure drop through one path with respect of Î¦$_T$. Furthermore, a phase diagram, illustrating the effects from operational parameter (Î¦$_T$), geometrical characteristics of distributor (C_d) and property of solids (Ga) on instability of uniformity, is attained to draw some useful conclusions of robust uniform fluidization.

509309 Numerical Simulation of Particle Sedimentation Related to Nuclear Safety By CFD-DEM Algorithm

Abstract:

Numerical Simulation of Particle Sedimentation related to Nuclear Safety by CFD-DEM Algorithm

Byoungcheol Hwang[1], Kiyofumi Moriyama[1], and Hyun Sun Park[1*]

[1]Division of Advanced Nuclear Engineering (DANE), POSTECH, Pohang, Korea

**corresponding author: hejsunny@postech.ac.kr*

Abstract

The Discrete Element Method (DEM) is a technique to treat particle-particle interactions, and is used in many industries such as civil engineering, powder processing, pharmacy, and so on. Coupling of this method with Computational Fluid Dynamics (CFD) is needed in problems where the particles interacts with fluid significantly, such as bubble column reactor, mineral processing, and so on.

The simulation of important phenomena in nuclear power plant accidents that involves significant core damage and melting, i.e. so called severe accidents, is also one of the topic where the CFD-DEM is useful. After the Fukushima accident in 2011, the severe accident analysis and preparation of counter measures to mitigate the accident consequences became regulatory requirements in many countries. One of the measures for the cooling and stabilization of the molten core is the flooding of the cavity under the reactor vessel where the dropping core melt is caught in a water pool. It is expected that the molten core breaks up into particles of several millimeter sizes and accumulates on the floor, then makes a particle bed, which is hopefully coolable. In this situation, the assessment for the coolability of the particle bed is very important, and a number of experimental and analytical approached suggest that the internal/external structure of the particle bed affects strongly the coolability. The CFD-DEM is a promising candidate for an analysis method that can clarify such details of the particle bed formation.

We tested the applicability of the multiphase CFD-DEM code, an open-source developed by CFDEM®coupling in 2011, for the simulation of particle sedimentation and particle bed formation under water. Our focus in the present work is on the prediction of the particle falling characteristics and particle bed shape. Although the real condition during a severe accident includes very high temperature particles (>1000K) and boiling water (two-phase flow) in a large scale, we started with a simulation of much simple experiment cases without heat transfer or gas phase, with spherical particles only as the first step for the validation of the method. The experimental data with solid particles of various diameters (4 mm, 6 mm) and densities (3600, 6000, 7800 kg/m^3), particle release nozzle diameters (20, 30, 40 mm) were referred to. The results of numerical simulation were compared with the experimental data in terms of particle falling time, maximum particle dispersion angle, particle average falling velocity, particle bed distribution, and so on. The particle falling time was well predicted by the simulation with the maximum deviation 5%. The maximum particle dispersion angle by the simulation significantly deviated from the experimental data probably due to the influences of initial disturbances. The particle average falling velocity was overestimated. The final particle bed profile by the simulation showed agreed with the experimental observation accurately in the peripheral region, but the height of the center part was underestimated. The fluid dynamic drag model and the particle-particle contact model would need improvement for better accuracy. The present work confirmed the validity of CFD-DEM algorithm on the application for the particle sedimentation in a simple liquid pool. The particle-pool interaction including the gas phase will be the next step.

Keywords: CFD-DEM; particle-fluid interaction; particle sedimentation; nuclear safety; severe accident

509440 Application of Extedned Discrete Element Method to the Melting Process of a Packed of Particles

Abstract:

Application of eXtedned Discrete Element Method to the melting process of a packed of particles

Mehdi Baniasadi, Maryam Baniasadi, Bernhard Peters, Xavier Besseron

Université du Luxembourg, Faculté des Sciences, de la Technologie et de la Communication, Luxembourg

e-mail: mehdi.baniasadi@uni.lu, maryam.baniasadi@uni.lu, bernhard.peters@uni.lu, xavier.besseron@uni.lu, web page: http:/www.xdem.de

ABSTRACT

The phenomena of melting takes place in many industrial processes, such as metal processing, environmental engineering and thermal energy storage systems. The problem of modeling of solid-liquid-gas flow including physical and chemical conversions of particles such as melting is the subject of intense research in industrial processes. These systems are difficult to model with numerical analysis due to the complex interaction between the different fluids and granular phases condering heat, mass and momentum excahnge. Basically models used to predict multiphase flows with particles are Eulerian-Eulerian and Lagrangian-Eulerian. The difference between the two models is based on different treatment of the movement of the particles. The particles are tracked individually in Lagrangian-Eulerian approach while they are considered as a continuum medium in Eulerian-Eulerian. Lagrangian-Eulerian is closer to the real physical processes rather than the Eulerian-Eulerian model for granular flows with particleâ€“particle interaction[1].

In this paper, our main focus is on the modelling of the melting process for a packed bed of particles subject to a fluid flow using a numerical technique based on Lagrangian-Eulerian so-called eXtended Discrete Elemnet Method (XDEM)[2]. In this approach, the particles are resolved as discrete phase coupled via heat, mass and momentum transfer to the surrounding fluid phase. The Discrete Element Method (DEM) coupled by heat transfer is used in order to evaluate temperature distribution, melting rate as well as velocity, oriantion and position for each particle individually. In addition the approach allows to be linked with continues numerical approaches such as Computational Fluid Dynamics (CFD) in order to solve fluid fields. The model has been validated by comparing predicted results with existing experimental data for melting of a single ice particle as well as a packed bed of ice particles in the horizontally flowing water. Fluid-particle and particle-particle intractions has also been considered. In addition, the model has the capability to be extended to the packed bed of particles with different size and properties to produce different liquid phases.

REFERENCES

[1] Dierich, F., Nikrityuk, P. A. & Ananiev, S. 2D modeling of moving particles with phase-change effect. *Chem. Eng. Sci.* **66,** 5459â€“5473 (2011).

[2] B. Peters, Thermal Conversion of Solid Fuels (WIT Press, Southampton, 2003).

514248 Real-Time Interpretation of Chaotic Characteristics and Instability of Dense Gas Solid Fluidization

Abstract:

The establishment of instability monitoring system targets at analyzing the chaotic characteristics and evaluating the instability of the fluidization process for dense gas solid fluidization. Experimental pressure fluctuation and pressure drop measurements were interpreted from dense gas solid fluidization system based on chaotic time-series analysis. The pressure fluctuation signals were filtered in terms of attenuating the noise signals. The filtered pressure signals were analyzed with respect to attractor reconstruction. Based on the reconstructed attractor, the maximum Lyapunov exponent and entropy were investigated and estimated under different bed heights and air velocities. Results show that the maximum Lyapunov exponents udner different bed heights and air velocities are positive, which indicates that dense gas solid fluidization exhibits chaotic characteristics. With the increase of bed height, estimated entropy of dense gas solid fluidization increases with the increase of bed height under identical air velocity, while it varies slightly with the increase of air velocity when the air velocity alters from minimum fluidization velocity to 1.36 times of the minimum fluidization velocity.

508510 Glucan Particles As Potential Carriers of Natural Flavonoids for Treatment of Idiopatic Inflammation Diseases

Abstract:

One of the novel methods of pharmaceutical formulation is the loading of the drug into biocompatible microparticles, preferably extracted from natural sources. This approach offers several potential advantages, *e.g.* increased bioavailability (the ratio of the administered drug, which actually enters the bloodstream) compared to conventional drug delivery methods, enabling the option of oral administration even for parenteral drugs. Multiple principles of the loading process are possible – drug encapsulation, adsorption onto the particle surface or precipitation inside the pores a hollow particle. The presented work focuses on precipitation of a drug from an organic solvent inside glucan particles. Glucan microparticles are ellipsoid porous shells, which are obtained from cell walls of *Saccharomyces cerevisiae* (baker´s yeast). The particle size is 2-5 micrometers. Cell walls of glucan particles are mainly composed of β-1,3-glucans. In order to prevent inflammatory reaction in organism, the yeast organelles are removed by a series of alkaline and acidic extraction steps. Subsequently, present lipopolysacharides are extracted using different organic solvents.
Lymphatic cells are mainly present in inflammation area in organism where they are responsible for progress of inflammation. Glucan particles (GPs) interact with lymphatic cell´s receptors on their surface and after that β-1,3-glucans are phagocytosed by the lymphatic cells. Therefore, GPs represent an effective oral delivery system for transporting of biological active substances. Curcumin is an anti-inflammatory flavonoid, which occurs in plants of family *Zingiberacaea.* In our project, curcumin was delivered to an inflammation in organism using GPs. Loading of curcumin into GPs was enabled via precipitation of curcumin from organic solvents. Curcumin dissolved in organic solvents was added to the GPs and through pores penetrated into cavity of glucan shell. Next, the solvent was removed and the curcumin was precipitated in glucan microparticles. GPs with the flavonoid inside were observed using confocal microscopy and SEM (scanning electron microscopy). The achieved content of curcumin was successfully analyzed by other methods, which can not be mentioned directly due to article in preparation. Furthermore, the curcumin-loaded GPs were tested *in vitro* on macrophage like cells and *in vivo* on laboratory mice in order to evaluate their effectivity. The obtained results show that these composites are promising materials for biomedical applications such as efficient treatment of inflammatory diseases (e.g. idiopatic inflammatory bowel disease, rheumatoic arthritis).

508601 Use of Advanced Imaging Techniques in Tablet Disintegration Study

Abstract:

Certified testing of tablet disintegration and dissolution provides us useful information. However, there are many more that are hidden due to techniques' restrictions. Such information is e.g. mechanism of disintegration, change of surface morphology or uniformity of the substances' distribution in the tablet. Therefore, there are many techniques used to obtain the remaining information such as Magnetic Resonance Imaging (MRI), Fourier Transform Infra-Red (FTIR) or Raman spectroscopy and microscopy, Terahertz spectroscopy, Scanning Electron Microscopy (SEM) and other.

Presented work uses tablet made by direct compression of binary mixture of ibuprofen and lactose (based on previous results there was 0.2 - 0.5 w/w of ibuprofen content used) pressed at different compaction pressures (low, medium and high). The tablets were subjected to several solutions and were studied by techniques that have insight into the tablet (MRI) and ones that can see the surface (FTIR and SEM).

MRI experiments provided us with several interesting and several expectable results. As the content of ibuprofen was increasing so was increasing the dissolution time and the wetted layer was decreasing. Similar results were observed when the compaction pressure was increased. When the tablets were submerged to dissolution limiting solutions (saturated lactose solution and solution saturated with both lactose and ibuprofen) the disintegration was severely limited and some tablets began to increase their volume. However, in case of 0.2 w/w ibuprofen tablets it didn't matter whether there was dissolution discriminating solution or not, it disintegrated anyway.

FTIR microscopy showed, that not only the amount, but also the shape and size of the particles from which the tablets were compacted can be determined. These findings were further supplemented by the dry state information obtained by Scanning Electron Microscopy after different dissolution times. Further the FTIR-ATR imaging technique helped to understand the basic mechanisms of the tablet disintegration and to study the dissolution and water penetration into the model ibuprofen-lactose tablets.

508632 Correlation Study between Liquid Penetration and Mechanical Properties of Pharmaceutical Tablets

Abstract:

Main phenomenon influencing treatment of patient in pharmacy is amount of active pharmaceutical ingredient dissolved in medium such as water or blood. Even though this isa key step to achieve desirable medical effect it is not the first one occurring.The first step is governed by surface properties such as wettability and roughness and the first contact point is followed by liquid penetration which is highly dependent not onlyon type of used liquid, ionstrength, liquid density but also on mechanical and structural properties of tablets such as porosity, structure strength or particle bonds.Through many analytical methods we can describe each of thesesteps individually. Aim of this study is to find proper correlation amongseveralmethods such as sessile drop method, texture analysis (TA) and magnetic resonance imaging (MRI) of tablet during dissolution.

For experiments binary mixture of tabletwas used. Tablets consist of lactose (hydrophilic) and ibuprofen (hydrophobic)with variable content (from 0.2 to 0.5 w/w ratio) and were also variable in compaction pressure used on them (from low â€“ 5 kN through 10kN as middle to 18 kN as high).Three different solutions were used according to their saturation with primary substances to prevent the tablet from dissolving. Surface properties were evaluated on contact angle and soaking rate of liquid, both measured by sessile drop method. Texture analysis was performed by recording the material resistance during dry and wet state of tablet. Then the data was processed and trends were obtained.Tablets during the MRI experiments were submerged into dissolution medium and the proceeding wetting front was measured. Later the images were processed via image analysis and trends were obtained.

Results obtained from surface experiments show that higher the ibuprofen content is the more hydrophobic surface appears to be and therefore the contact angle is increasing. Contact angle is independent on manufacturing pressure but soaking rate on the other hand depends strongly.MRI images show that this technique is not suitable for rapidly disintegration tablets, where the calculated error is within experiment conditions. On the other hand, tablet with prolonged dissolution are well measured. Results from MRI experiments also state that local saturation with substances supress dissolution of tablet. Texture analysis of tablets show that they undergo immediately structural weakening after being poured with solution independently on type of solution. Rate of this structural change mainly depends on solution's acidity. Thenthe tablet decreases its resistance in time which can be interpreted as relaxation rate. With increment of ibuprofen or saturation with primary substances this rate decrease, which means that dissolution is suppressed. Correlation between MRI and relaxation rate was done. Results shows that the correlation is mainly dependent on texture analysis measurement. On the other hand, correlation between MRI and soaking rate appears to be more tentative.

Experiment characterising surface, structural and liquid penetration properties were investigated by three different techniques. Obtained results were evaluated and put to mutual correlation. From liquid penetration point of interest there is correlation between MRI and contact angle method. Also, a correlation between MRI and texture analyser was found.

508642 In Situ Drug Amorphisation By Microwave Irradiation Stabilized By Mesoporous Silica

Abstract:

One of the common issues in the pharmaceutical industry is the poor solubility of the newly developed active pharmaceutical ingredients (APIs). The solubility might be significantly enhanced by changing the form of the API from crystalline to amorphous form. The amorphous state is characterized by lower thermodynamic stability and thus better dissolution properties leading to better bioavailability. In addition, the amorphous API in a dosage form (e.g. tablets) must also remain stable and not recrystallize for at least the shelf life of the drug product (usually 2 - 3 years). Various methods are being utilized to prepare and stabilize amorphous formulations in the industry, however they usually involve difficult formulation steps or suffer from poor efficiency, low stability or can be patent obstructed.

In this work a stable amorphous formulation is prepared simply using the standard wet and dry granulation process or direct tableting. Crystalline form of API is used during the initial process. The amorphous form is then prepared *in situ* and â€œon demandâ€ either during manufacturing, or by the end user by applying microwave radiation (MW).

To stabilize the API in an amorphous form after the MW, multi-structured mesoporous silica particles were added to the formulation. The particles were prepared using an emulsion synthesis of octylamine in tetraethyl orthosilicate. Using ethanol as an inhibitor instead of the commonly used nitric acid as a catalyst reduces the silica primary nano-particles nucleation rate. The formed nuclei grow larger and then self-assemble around the octylamine droplets and form hollow micro-particles. The prepared microparticles were characterized using the Scanning electron microscopy (SEM) and the Transition electron microscopy (TEM). To characterize the porous structure, nitrogen sorption (BET) analysis was conducted. The results clearly show the particle assemblies contain macropores (cavities), mesopores and micropores and possess very high specific surface area and also very large pore volume. Together with the thermal and chemical stability of silica these properties make the particles well suitable for preserving the melted API in an amorphous form after the MW.

As a proof of concept, initially powder mixtures and later simple tablets composed of Ibuprofen (model API), silica micro-particles and a filler excipient were experimentally investigated. The *in situ* amorphization of ibuprofen was conducted using a commercial microwave oven. The heating properties of the oven, i.e. the spatial distribution of microwave heating was characterized. Various microwave power input and varying composition of the mixtures (excipient/Ibuprofen ratio) were tested with respect to the melting point of the Ibuprofen. The change from the crystalline to amorphous structure due to MW heating was analyzed by x-ray diffraction (XRD), differential scanning calorimetry (DSC) and FTIR-ATR techniques. The dissolution profiles of ibuprofen from both MW untreated and treated tablets were measured using the High Performance Liquid Chromatography (HPLC).

Based on these initial results, fully formulated tablets were prepared with various API dosage strength using a tablet press. Same compression force and same dwell time was used for all of the tablets. The final tablet composition includes excipient capable of absorption of the MW, which ensures a uniform heating in the whole volume of the tablet. Next it contains the silica microparticles, which by capillary forces absorb the API once melted and stabilize it after cooling down.

Lastly, the tablets contain standard filler, lubricant and disintegrant excipients. After the MW application the tablets change their structure due to the melted API intake into the silica micro-particles, forming visible pores in the tablet structure. This has a notable impact on the dissolution and hardness of the resulting tablets. For the final tablet characterization SEM, Texture Analysis, FTIR and Magnetic Resonance Imaging were used.

508646 Manufacture of Personalised Medicines By API Printing on Porous Tablets

Abstract:

Personalised medicine offers an opportunity to produce tailor-made medicaments for patients. In order to realize this vision, it is necessary to find an alternative approach to the conventional tablet design and production. One that will replace the existing mass production of drugs offering only a few dose options of the active pharmaceutical ingredient (API), while at the same time it must meet the strict reproducibility and a defined relationship between composition and release rate of active substances requirements.

One of such approaches is to print the API on demand on a solid substrate directly when it becomes needed. Appropriate dosing of API and the possibility to even combine several APIs into a single medicament improves the variability of the dosages available for prescription and reduces the overall number of pills the patient has to take. The goal of this work is to find a simple, automated and cost-effective method of preparing personalised medicaments, which could take place directly in the pharmacy shops or at home.

The method in this work consists of a precise dosing of a dissolved API into an already prepared substrate â€“ porous placebo tablets with high sorption capacity. The tablets are prepared using direct compression in a tablet press from a simple physical mixture of a binder, filler, disintegrant and lubricant. As a filler, mesoporous silica particles were used. The high pore volume provides the free space for API sorption into the tablets and the small size of the mesopores ensure the API is deposited in an amorphous form.

The prepared particles used in the formulation were characterized using Transmission electron microscopy (TEM), Dynamic light scattering (DLS) and nitrogen sorption analysis (BET). The prepared tablets hardness and uniformity was characterized both prior and after loading. The tablets after loading were characterized using the X-ray powder diffraction and the Fourier transform infrared spectroscopy to determine the API form and deposition uniformity, followed by dissolution tests to confirm the applied dosage strength and to measure the release profile. So far, several BCS Class II APIs were successfully infused into the substrates using precise on-demand dosing. Impregnation of various doses of API were accurate. Later characterization revealed no crystalline API were present and the dissolution profiles were very satisfactory.

509441 Understanding Phase Transition of Acetaminophen in the Bulk and Surface of Acetaminophen

Abstract:

Pharmaceutical solids processing often lead to the generation of new phases in the bulk and on the surface of the processed drugs. Phase transformation of the API can potentially change the physiochemical properties and influence the drug performances including solubility, dissolution and stability. To detect these changes, analytical tools such as XRD are usually employed. However, these techniques remain suitable for bulk analysis. Whereas, phase transition on the surface remains hard to detect due to the lack of literature, researches in this area and the absence of surface analytical techniques. The aim of this work is to investigate different phases (pure crystalline, partially crystalline partially amorphous and amorphous) of Acetaminophen drug by combining bulk and surface analytical techniques. For bulk analysis, FTIR and XRD have been employed. The former shows no significant differences between Acetaminophen various phases. Whereas the latter confirmed the presence of different phases indicated by a remarkable variation in the absorbance intensity the peak width. Where, high and sharp peaks obtained for the pure crystalline Acetaminophen, fewer peaks observed for semi crystalline Acetaminophen and a single broad peak observed for the amorphous Acetaminophen. For the surface analysis, Nitrogen K near-edge X-ray absorption fine structure (NEXAFS) spectra of a series of NEXAFS indicates that the precise determination of IPs and Ï*f** shape resonance energies is sensitive enough to distinguish between the different phases of Acetaminophen. DFT calculations provides information on the molecular orbitals involved during electronic excitation occurred as a result of NEXAFS experiment.

509599 Treatment of Cystinosis through Vitamin E Modified Silicone Hydrogel

Abstract:

Cystinosis is an orphan disease caused by a genetic mutation that leads to deposition of cystine crystals in many organs including cornea. Ophthalmic manifestation of the disease can be treated with hourly instillation of cysteamine eye drops. The hourly eye drop instillation is tedious to the patients leading to poor compliance and additionally, significant degradation of the drug occurs within one week of opening the bottle, which further complicates this delivery approach. Our work focuses on designing a contact lens to treat the disease with improved efficacy compared to eye drops. Our goal is to design a lens that is safe and that can deliver a daily therapeutic dose of cysteamine to the cornea while retaining drug stability. We show that cysteamine diffuses out rapidly from all lenses due to its small size. The rapid release of cysteamine and other ophthalmic drugs can be prolonged significantly by integration of vitamin E aggregates inside of the polymer matrix of silicone hydrogels. The increase in release duration increases as square of the volume fraction of the loaded vitamin E fraction which leads to high efficacy and suggests that the mechanism of attenuation in drug release is formation of nanosized barriers. The increase in duration depends on the type of the contact lens. The barriers are not effective in single-phase hydrogel lenses while the effect in the bi-phasic silicone hydrogels depends on the type of the contact lens. Vitamin E incorporation increases the release duration of both ACUVUE®OASYS® and ACUVUE® TruEye™ but the effect is more pronounced in TruEye™ likely due to the low solubility of vitamin E in the lens matrix and higher aspect ratio of the barriers. This result along with the high aspect ratio seen through SEM imaging in silicone hydrogels suggests that barriers could be forming at the interface of the silicone and hydrogel phases. The presence of vitamin E has an additional beneficial effect of reduction in the oxidation rates, likely due to a transport barrier between the oxygen diffusing through the silicone channels and drug located in the hydrogel phase. Based on this study, both Acuvue®OASYS® and ACUVUE® TruEye™ can be loaded with vitamin E to design a cysteamine eluting contact lenses for effective therapy of cystinosis. The lenses must be worn for about 4-5 hr. each day, which is less than the typical duration of daily-wear. Vitamin E loaded contact lenses could be very useful for treating many other ophthalmic diseases as well.

517683 Systems Integration for Dry Granulation Based Continuous Pharmaceutical Tableting

Abstract:

Over the past decade, the pharmaceutical industry has seen significant progress in the mechanistic understanding of relevant unit operations, availability of multiple sensor technologies for inline monitoring of critical quality attributes and advances in data analytics to support manufacturing decision-making. These have enabled a systematic realization of continuous manufacturing for pharmaceutical tablets using the direct compaction route. However, the limitations and difficulty in handling of some particulate materials provide the impetus for investigating granulation routes to the continuous manufacturing of tablets.

In this work, the integration of dry granulation into a continuous tableting line is investigated. Specifically, the Process Analytical Technology (PAT) tools required for inline and at-line monitoring of the tableting line, and the semi-mechanistic models for the roller compactor and the tablet press are discussed. While the Kawakita model is used for the tablet press, the roller compaction process model is consisting of an integration of material balance and actuator dynamics from Hsu et. al (2010) with the modified Johansonâ€™s model presented by Liu & Wassgren (2016). The PAT tools, mechanistic equations, and automation system are integrated for measurement validation and data accuracy using data reconciliation in a structured ISA-95 network using DeltaV. The experimental data is presented from the operation of the continuous dry granulation line-based pilot plant at Purdue University.

509248 Modeling Granular Material Mixing and Segregation Using a Multi-Scale Model

Abstract:

A multi-scale model is presented for predicting the magnitude and rate of powder mixing and segregation in blenders. The model combines diffusion and percolation correlations from small-scale discrete element method (DEM) simulations with advective flow field information from large-scale finite element method simulations. The multi-scale model predictions closely match results from both large-scale DEM-only simulations and experiments, but they are much more computationally efficient than DEM-only simulations. Therefore, in contrast to DEM, the multi-scale model is well suited for predicting blending in industrial-scale systems.

Modelling Granular Media using Dynamical Density Functional Theory

T. Hurst, B. Goddard & R. Ocone

Motivation

Granular media is ubiquitous in industrial and physical problems. One of the characteristic qualities of granular media at the microscopic level is inelastic hard body collisions, which drive macroscopic behaviour. Modelling granular media is a complex and interesting problem. Solid-like phenomena make it difficult to approximate a granular system with a continuum model, while chaotic behaviour and inelastic collapse in hard particle systems introduce modelling challenges on the microscopic scale. We investigate uses of dynamical density functional theory (DDFT), augmented with data gained from microscopic models, for granular media simulation.

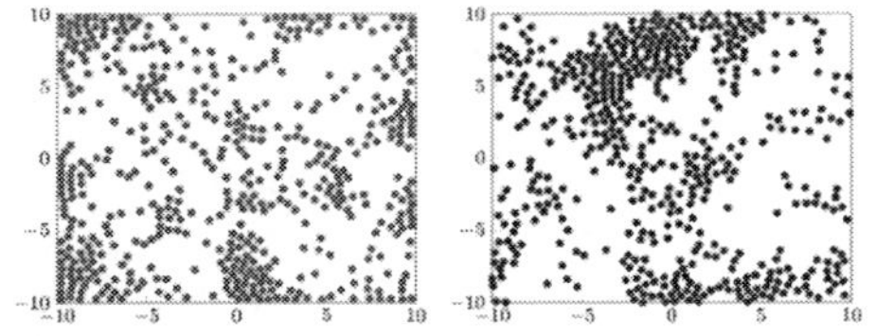

Figure 1: Inelastic systems with reflective (left) and periodic (right) boundaries

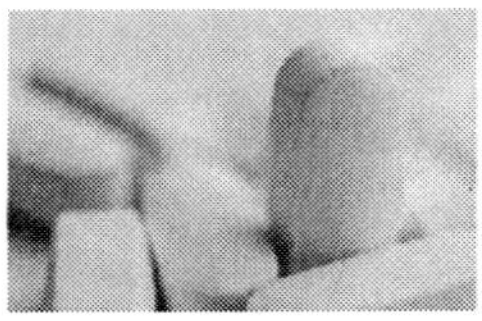

Dynamical Density Functional Theory

To derive a DDFT, we start with a set of Langevin equations for N hard particles of diameter σ in d dimensions with positions $\boldsymbol{r}^N$ and momenta $\boldsymbol{p}^N$:

$$\frac{d\boldsymbol{r}^N}{dt} = \frac{\boldsymbol{p}^N}{m}, \quad \frac{d\boldsymbol{p}^N}{dt} = -\nabla_{\boldsymbol{r}^N} V(\boldsymbol{r}^N, t) - \gamma \boldsymbol{p}^N + (mk_bT\gamma)^{1/2}\boldsymbol{G}(t), \tag{1}$$

on the hard sphere domain, $\mathcal{D}_N = \{(\boldsymbol{r}^N, \boldsymbol{p}^N) \in \mathbb{R}^{2dN} : \forall i,j, \|\boldsymbol{r}_i - \boldsymbol{r}_j\| \geq \sigma\}$, where $V(\boldsymbol{r}^N, t)$ includes external and interparticle potentials, and $\boldsymbol{G}(t)$ is an optional stochastic force term. The Kramer's equation is then a partial differential equation for its distribution function $f^{(N)}(\boldsymbol{r}^N, \boldsymbol{p}^N, t)$:

$$\frac{\partial f^{(N)}}{\partial t} + \frac{1}{m}\boldsymbol{p}^N \cdot \nabla_{\boldsymbol{r}^N} f^{(N)} + \nabla_{\boldsymbol{r}^N} V \cdot \nabla_{\boldsymbol{p}^N} f^{(N)} = \gamma \nabla_{\boldsymbol{p}^N} \cdot \boldsymbol{p}^N f^{(N)} + \gamma k_b T \nabla^2_{\boldsymbol{p}^N} f^{(N)}. \tag{2}$$

We integrate out all but one position and momenta in the Kramers equation, to find:

$$\frac{\partial}{\partial t} f^{(1)} = \mathcal{L}_1 f^{(1)} + \mathcal{L}^{2,\dots,N}_{\text{part}}(f^{(k)}) + \mathcal{L}^{(1)}_{\text{diss}} f^{(1)} + \mathcal{L}_{\text{coll}}(f^{(2)}), \tag{3}$$

where $\mathcal{L}_1$ introduces effects from external potentials, $\mathcal{L}^{(1)}_{\text{diss}}$ effects due to dissipation and friction, $\mathcal{L}^{2,\dots,N}_{\text{part}}$ effects from interparticle potentials, and

$$\mathcal{L}_{\text{coll}}(f^{(2)}) = -\sigma^{d-1} \int_{\mathbb{R}^d} d\boldsymbol{p}_1 \int_{\mathbb{R}^d} d\boldsymbol{p}_2 \int_{S_1^{d-1}} (\boldsymbol{p}_1 - \boldsymbol{p}_2) \cdot \omega f^{(2)}(\boldsymbol{r}_1, \boldsymbol{r}_1 + \sigma\omega, \boldsymbol{p}_1, \boldsymbol{p}_2, t) d\omega, \tag{4}$$

is a general collision term that arises due to the hard-sphere domain, when assuming only binary collisions. By taking the first two $\boldsymbol{p}$-moments and upon additional assumptions we arrive at a closed set of equations to model the dynamics of the system, a DDFT [1, 2]:

$$\frac{\partial}{\partial t}\rho = -\nabla \cdot (\rho \boldsymbol{v}), \tag{5}$$

$$\left(\frac{\partial \boldsymbol{v}}{\partial t} + \boldsymbol{v} \cdot \nabla \boldsymbol{v}\right) + \gamma \boldsymbol{v} = -\frac{1}{m}\nabla \frac{\delta \mathcal{F}[\rho]}{\delta \rho} - \sigma^{d-1} g_\sigma \int_{S_1^{d-1}} \rho(\boldsymbol{r} + \sigma\boldsymbol{\omega}/2)\rho(\boldsymbol{r} - \sigma\boldsymbol{\omega}/2)$$
$$\times \boldsymbol{\omega} \cdot \left\{ \frac{k_B T}{m}\mathbf{1} + \boldsymbol{v}(\boldsymbol{r} + \sigma\boldsymbol{\omega}/2) \otimes \boldsymbol{v}(\boldsymbol{r} + \sigma\boldsymbol{\omega}) - \boldsymbol{v}(\boldsymbol{r} + \sigma\boldsymbol{\omega}/2) \otimes \boldsymbol{v}(\boldsymbol{r} - \sigma\boldsymbol{\omega}/2) \right\} \mathrm{d}\boldsymbol{\omega}, \tag{6}$$

where $\mathcal{F}$ is the well-studied Helmholtz free energy operator, which can include particle-particle interactions and external potentials, ρ is the density, $\boldsymbol{v}$ is the local velocity and g_σ is the value of the correlation function at σ, which must be determined. We choose to model g using statistical data from microscopic simulation.

DDFT Simulations

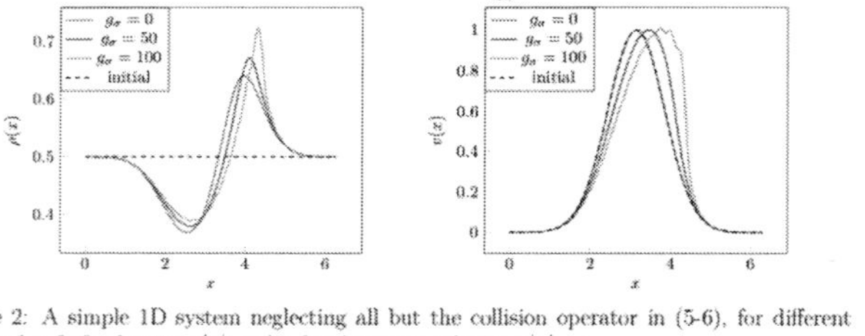

Figure 2: A simple 1D system neglecting all but the collision operator in (5-6), for different values of g. Left: density $\rho(x)$, right: local average velocity $v(x)$.

Data-Driven Modelling of g

A system of N hard spheres of diameter σ in d dimensions with positions $\boldsymbol{x}^N$ and velocities $\boldsymbol{v}^N$ obey

$$\frac{d\boldsymbol{r}^N}{dt} = \boldsymbol{p}^N, \quad \frac{d\boldsymbol{p}^N}{dt} = V(\boldsymbol{r}^N, t), \tag{7}$$

with a pairwise collision condition: if particles i and j are in contact, i.e. $\|\boldsymbol{r}_i - \boldsymbol{r}_j\| = \sigma$ and $(\boldsymbol{r}_i - \boldsymbol{r}_j) \cdot (\boldsymbol{p}_i - \boldsymbol{p}_j) < 0$, we instantaneously change the velocities via the rule:

$$\boldsymbol{p}_i^{out} = \boldsymbol{p}_i^{in} - \frac{1+\alpha}{2}\boldsymbol{\nu}_{ij} \cdot (\boldsymbol{p}_i^{in} - \boldsymbol{p}_j^{in})\boldsymbol{\nu}_{ij}, \quad \boldsymbol{p}_j^{out} = \boldsymbol{p}_j^{in} + \frac{1+\alpha}{2}\boldsymbol{\eta}_{ij} \cdot (\boldsymbol{p}_i^{in} - \boldsymbol{p}_j^{in})\boldsymbol{\eta}_{ij}, \tag{8}$$

where $\boldsymbol{\eta}_{ij} = \frac{\boldsymbol{r}_i - \boldsymbol{r}_j}{\|\boldsymbol{r}_i - \boldsymbol{r}_j\|}$ and $\alpha \in [0, 1]$ is the coefficient of restitution. For simple V, the Newton equations can be modelled using Event Driven Particle Dynamics (EDPD). The cell method and bounded increasing priority queues [3] reduce the computational effort to $\mathcal{O}(1)$ per collision. Using statistical methods for synthetic data (e.g. emulation [4]) we gain an understanding of the correlation function g near equilibrium, which we can include in (6-7).

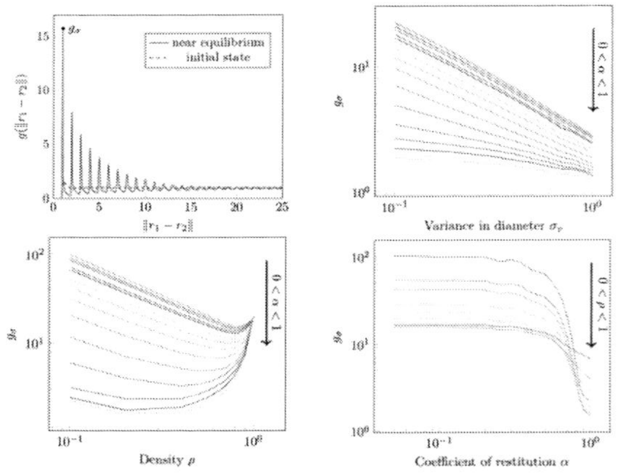

Figure 3: Top left: $g(\|r_1 - r_2\|)$ near equilibrium for a 1D system with $\rho = 0.5$ and $\alpha = 0.5$, using 10000 samples of EDPD simulations. The other three graphs show how g_σ changes for different parameters.

Future Work

We wish to include additional moments in the DDFT derivation, simulate DDFT in higher dimensions including interparticle potentials, and use EDPD in higher dimensions to calculate g.

[1] Archer, A., *J.Chem.Phys.*, **130**, 014509, (2009).
[2] Goddard, B. et. al., *Phys.Rev.Lett*, **109**, 1-5, (2009).
[3] Paul, G., *J.Comp.Phys.*, **221**, 615-625, (2011).
[4] O'Hagan, A., *Reliab.Eng.Syst.Saf.*, **91**, 1290-1300, (2006)

The support from EPSRC (grants EP/N034066/1, EP/L025159/1 and EP/L016508/01) is kindly acknowledged.

509555 A Study on Partially Wetted Particle Collisions with a Wet Wall

Abstract:

For the production of poly-ethylene, a gas-phase polymerization reaction is executed in a fluidized bed reactor. The polymerization reaction produces an excess of heat and the reactor is unable to efficiently remove the excess heat. The introduction of an Inert Cooling Agent (ICA) solves the temperature problem but also has a downside. Introducing the ICA in the reactor influences the particle collisions, leading to substantially different flow behavior. Besides the reduction of the restitution coefficient, the introduction of the liquid can result in adhesion instead of collision. This adhesion results in the formation agglomerates in the reactor and reduces the efficiency of the process. Due to the complex physics of this problem, Computational Fluid Dynamics (CFD) is a suitable method to gain insight in this phenomenon.

In order to study this complex system of partially wetted particle collisions a Direct Numerical Simulation (DNS) is used for incompressible flow. To model the coupling between the solid phases and the fluid phase, a second order accurate Immersed Boundary Method (IBM) is used (Deen et al.,2012). The interface between the fluids is tracked by a Volume Of Fluid (VOF) method (Baltussen et al., 2017). The contact-line dynamics are solved by imposing a contact angle as a boundary condition (Patel et al., 2017).

The simulated results will be compared with dry particle collisions on a wet wall (Tang et al., 2017) and with wet particle collisions on a wet wall.

References:

1. Deen, N.G., Kriebitzsch, S.H., van der Hoef, M.A. and Kuipers, J.A.M., Direct numerical simulation of flow and heat transfer in dense fluidâ€"particle systems. Chemical Engineering Science, 81 (2012): 329-344.

2. Baltussen, M. W., Segers, Q. I. E., Kuipers, J. A. M., and Deen, N. G., Cutting bubbles with a single wire. Chemical Engineering Science 157 (2017): 138-146.

3. Patel, H.V., Das, S., Kuipers, J.A.M., Padding, J.T. and Peters, E.A.J.F., A coupled Volume of Fluid and Immersed Boundary Method for simulating 3D multiphase flows with contact line dynamics in complex geometries. Chemical Engineering Science, 166 (2017): 28-41.

4. Tang, Y., Buck, B., Heinrich, S., Deen, N.G., Kuipers, J.A.M., Interface-resolved simulations of normal collisions of spheres on a wet surface. American Institute of Chemical Engineers, (2017).

514089 The Vagaries of Granular and Particle Laden Flow

Abstract:

In the 80s, the development of the kinetic-collisional theory for granular flow (KTGF) represented a major breakthrough in the modelling of the hydrodynamics of particulate flow: based on the analogy between solid particles and gas molecules, a kinetic theory was proposed which produced constitutive equations for rapid granular flow when binary, instantaneous and inelastic particle collisions occur. In practical situations, however, the flow of particles presents complexities that escape successful description based on the KTGF alone. Many complicated phenomena happen at the meso-scale, the scale between "small" and "large": as an example, clusters and agglomerates are formed at a length scale between particle size and equipment size. In addition, particle-particle interactions, other than inelastic and instantaneous collisions, might occur, and more complex theories are needed to capture the observed behaviour.

The talk will describe the journey from the first rigorous effort to model granular and particle laden flow to recent theories and models to explain the vagaries observed in practical situations. Open problems, in need of a solution, will be presented and the search for a "unifying" theory will be discussed.

514090 Scaling Up Particulate Process - Lessons from Industry

Abstract:

In the 1980â€™s, the Rand Corporation identified challenges in the commercialization of processes involving particles. Particulate process plants took longer to start up and achieved lower production rates than expected. Not enough has changed in more than 30 years. Processes involving particles are inherently more difficult than those for liquids and gases. A consistent lack of fundamental understanding dooms industry to empirical approaches, and the sensitivity of experiments to scale and equipment configuration leads to high costs and long development periods. Modeling by the discrete element method (DEM) has not yet fulfilled its promise, especially for small particles. This presentation will provide examples of the challenges involved in some of the unit operations of particle technology, and discuss strategies for scale-up, including the use of equipment suppliers, toll processors, modeling and pilot plants to bridge the knowledge gaps. Even with such strategies, education of both the engineer and the business leader is necessary to manage expectations and appropriately allocate resources.

507320 Experimental and Numerical Study of Structure and Hydrodynamics in Packed Beds of Spherical Particles

Abstract:

In chemical industry, flows are often multiphase and occur in non-transparent equipment, for example in steel pipelines and vessels. MRI is a suitable approach to visualize and quantify the flow, that cannot be seen with classical optical techniques. It is therefore a unique tool to non-invasively obtain whole-field porosity and velocity distributions in opaque dispersed multiphase flows.

One of the most common chemical reactors is the packed bed, in which the fluid phase flows through a fixed bed of solid particles. In many applications the solid particles are catalyst pellets. This reactor is usually operated with a wide range of conditions, and its performance is strictly related to the flow fields of the fluid. Before the introduction of tomography and MRI, only overall properties of a packed bed could be determined, such as the fluid flow rates, pressure drop, liquid holdup, wetting efficiency, axial dispersion coefficient and residence time distribution. Even though global properties of the hydrodynamics are useful, they are unable to provide a complete overview of the performance of the reactor, since the inhomogeneity of the packing structure causes significant local variations in holdup, wetting and fluid velocities.

Apart from imaging the fluid distribution in the bed, MRI allows non-invasive measurements of the local fluid velocities. These data can be used to check for the accuracy and appropriateness of assumptions used in numerical modelling of packed beds.

In the present investigation, experimental results obtained with this technique, applied to the study of structure and hydrodynamics in packed beds of spherical particles are shown and compared with CFD simulations performed with an in-house numerical code based on an Immersed Boundary Model-Direct Numerical Simulation (IBM-DNS) approach.

Pressure drop and radial profiles (averaged along the axial direction) of porosity and axial velocity of the fluid were determined, by comparing experiments with simulations, for packed beds of spherical particles with different sizes.

Particular attention was paid to the values of experimental results when directly encoding velocity by means of MRI, since the overall material balance, which relates the average bed porosity with superficial and axial velocity of the fluid, must always be satisfied.

Structure, in terms of phase fractions and, therefore, local porosity, was reconstructed through an algorithm able to detect spheres in a stack of slices. Experimental 3-D datasets of the packed beds have been obtained with static MRI measurements, whereas the axial velocity of water flowing through the particles was obtained by directly encoding velocity values.

From the axial velocity map, a PDF of experimental velocity values was also obtained for each sphere size and flow condition, showing good agreement with the overall material balance involving the relationship among average porosity, superficial and average axial velocity of the fluid.

Since the IBM-DNS method was developed and used to perform CFD simulations of transport phenomena in packed beds only in recent years, this work tried to fill the gap related to validation of the numerical approach, by means of a one-to-one comparison with detailed 3-D datasets obtained through MRI.

507348 Analyzing Mechanical Labile Protein Crystals with a Special Filtration Cuvette

Abstract:

BACKGROUND: Recent progresses in the biotechnology industries have led to much higher product yields after the fermentation. In the case of proteins the purification of such high titers have become a challenging task. The common method is to use several chromatography steps to separate the protein with high purity. However the high concentration of protein in the mother liquor requires large amounts of toxic and explosive organic solvents and large chromatography columns. Hence another approach like selective crystallization to separate the protein from the mother liquor is becoming increasingly interesting for the pharmaceutical and biotechnology industry. Crystallized proteins not only offer advantages in the downstream processing but also have other positive properties like extended shelf life, easier product handling and in the case of therapeutic proteins different drug release properties.

After the crystallization the next step is often a solid-liquid separation. In the case of protein crystals this separation is a challenging task because in contrast to common organic or inorganic crystals protein crystals are much more fragile to mechanical stress. Even low pressure differences may lead to crystal breakage or crystal abrasion. Since therapeutic or enzymatic proteins are often expensive products a method to analyze the mechanical stability and the influence of crystal breakage on the cake filtration with low volumes and hence low product quantities is required.

METHODS: To analyze the influence of crystal breakage on the cake filtration a special filtration cuvette has been constructed which allows the filtration of volumes less than one milliliter. This 3D-printed cuvette can be used in an optical, analytical centrifuge by LUM GmbH (Berlin, Germany) to perform and monitor the filtration by means of light transmission at the same time. As a model substance hen egg-white lysozyme crystals were investigated. The particle size distribution can be measured by automatic analysis of microscope images or by analysis of sedimentation velocity. By comparing the particle size distributions before and after the filtration the crystal breakage can be quantified at different centrifugal accelerations. Due to the light transmission unit in the centrifuge the filter cake and the flux through the filter can be observed during the filtration. Different crystal shapes and sizes of lysozyme were investigated using this method.

OBJECTIVE: The objective was to analyze the mechanical stability of different shapes of lysozyme crystals during the filtration in the centrifugal field. Therefore the crystals were stressed at different accelerations in the Lumisizer® using a special 3D-printed filtration cuvette. The experimental setup and the results of experiments with different crystal shapes of lysozyme will be presented in detail.

RESULTS: Due to crystal breakage during the filtration the porosity of the filter cake decreases leading to a lower flux through the filter cake. Especially aggregates may break leading to a higher fines content and reducing the coarse content. Hence the unstressed crystal size distribution of aggregated lysozyme shows a larger coarse content and a larger median particle diameter than the crystals which were stressed at for instance 2000 or 4000rpm. Experiments with isometric and rod-like crystals have also been performed to compare the amount of mechanical stress and the reduction in the crystal size distribution. Using the filtration cuvette the filtration progess, the filter cake consolidation and the flux through the filter can be compared between the different particle systems stressed at different centrifugal forces.

Up scaling the IsaMill with the enhanced stress model

S. Breitung-Faes, D. Schons, A. Kwade

Institute for Particle Technology, TU Braunschweig

The IsaMill is an often used type of wet operated stirred media mill in the mining and minerals industry.Grinding chambers with volumes up to 10 m³ are generally used enabling the required production rates in the industrial scale.. The optimum set-up of such a grinding process, especially the choice of process parameters is often connected to empirical approaches or rough scale-up rules. In the mining industry e so called signature plots are widely spread, whereas in general stirred media milling processes are also transferred with length to diameter ratios.. A more scientific way for scaling is the usage of experiments in a small scale find optimum grinding media diameters and stirrer tip speeds. These parameters can be integrated into the so called stress energy, which is comparable to the kinetic energy of colliding grinding media [1, 2]. The stress energy is a value where the most important grinding parameters are introduces. Thus for each particle size and type an optimum stress energy can be determined by plotting the required energy demands over the applied stress energies.If the process parameters are set to values that the optimum stress energy occurs and the power consumption of the mill is at a high level, maximum production rates result. Each mill has a stress energy distribution. Its median value is a characteristic value for a mill. It should be kept constant for transferring or scaling a process by adopting the process parameters.

For nanogrinding in stirred media mills this origin idea of using the stress energy was used to establish an enhanced stress model, which enables the prediction of optimum stress conditions for different mill designs, sizes and materials on basis of geometrical data and material values as input parameters [3-6]. Therefore the median stress energy is used for different mill sizes and geometries, which can be calculated by geometrical data of the mill. Besides the prediction of the optimum grinding conditions with less experimental work the prediction of specific energy input for a given set of parameters and a determined product quality can be realized. This model was used for lab scale mills (grinding chamber volumes below 2 L) and especially for nano applications where the stability and rheology of the suspensions plays a major role. The connection of material, process and geometry parameters lead to two approaches of using the model:

- Determination of optimum process parameters for known materials for desired product qualities in any mill geometry of size
- Determination of material parameters from experimental milling results.

Within this work the model is applied for both approaches. First experimental results obtained in a M4 IsaMill were used to determine the material parameters of a dolomite. Afterwards the enhanced stress model is used to design a grinding process of this material in a M20 pilot scale IsaMill. Results obtained with the enhanced stress model were compared with the results obtained by using the origin optimization way with the stress energy curves and Power number/Reynolds number curves [7]. It can be concluded that the predictions of the model are in a good range in comparison to the experimental data and adoptable for various sets of parameters. Therefore, only a small experimental effort is necessary, much smaller than for the classical optimization and set up route.

Material and Methods

For the determination of the materials parameters and the scale up of the grinding process, two different IsaMills were operated in circuit mode. The lab scale M4 IsaMill (5.5 kW) was used for the determination of the product material parameters and the pilot scale M20 IsaMill (18.5 kW) was used to verify the validity for the scale up procedure by the enhanced stress model The power consumptions of the mills were recorded by torque meters. A constant suspension temperature was realized by installing a heat exchanger into the grinding circuit.

As grinding media material alumina beads of the type Microbit Leonardo from Industrie Bitossi (Italy)
were used in two different sizes: 1.5 mm and 2.5 mm. The filling ratio of the grinding media inside the grinding chambers was kept constant at 70 %.

As feed material a dolomite provided also by Industrie Bitossi was used. It was ground as an aqueous suspension with a solids content of 20 vol.%. The particle size analysis was carried out using a Mastersizer 3000 (Malvern Instruments). The maximum initial particle size was located in the range of 120 µm.

Introduction to the enhanced stress model

The original stress model was developed by Kwade [2, 8]. He showed that for wet stirred media milling the relation between specific energy and stress energy leads to minimum curves if the particle size is kept constant. The stress energy reflects therewith more or less the kinetic energy of colliding grinding media and is expressed with the stirrer tip speed v_t, the grinding media size d_{GM} and density ρ_{GM}. For the enhanced stress model Breitung-Faes et al. included a mill specific factor f_{mill} which recognizes the mill geometry [9], thus the following equation occurs:

$$\overline{SE}_{GM} = f_{mill} \cdot v_t^2 \cdot d_{GM}^3 \cdot \rho_{GM} \qquad \text{Equation 1}$$

Besides the energy, which is provided by colliding grinding media the stress number or number of stress events during a certain grinding time influences the grinding result. The stress number cannot be measured and has to be estimated by the operation parameters in a similar way as the stress energy. Thus, the stress number SN_{GM} was defined as product of stirrer revolution number $n_{stirrer}$, number of grinding media inside the grinding chamber N_{GM} and grinding time t_{GM}. The grinding media number can be expressed via the total mass of grinding media filling m_{GM} as well as grinding media diameter d_{GM} and density ρ_{GM}. Depending on the mill geometry a constant which is a function of the grinding media diameter has to be introduced, because experimental results showed, that the power consumption of a mill not only depends on the total grinding media mass and stirrer tip speed, but in some cases also on the grinding media diameter, especially if dynamic separation systems (like classifier wheels) for the grinding media are used.

$$SN_{GM} = k(d_{GM}) \cdot SF \cdot t_g = k(d_{GM}) \cdot n_{Stirrer} \cdot \frac{m_{GM}}{\frac{\pi}{6} \cdot d_{GM}^3 \cdot \rho_{GM}} \cdot t_g \qquad \text{Equation 2}$$

The product of stress energy $\overline{SE}_{GM}$ and stress number SN_{GM} is proportional to the energy consumption of the mill.

$$E \sim \overline{SE}_{GM} \cdot SN_{GM} \qquad \text{Equation 3}$$

The enhancement of the original stress model was realized in two steps. The first step was focused onthe introduction of breakage relevant material parameters to predict optimum stress energies for desired product fineness. The second step was dealing with the determination of the grinding times or specific energy for desired product fineness and chosen process parameters. Here both parts are shortly summarized, because both parts are necessary for the scale-up of the process.

Calculation of optimum stress conditions with the enhanced stress model

The stress energy $\overline{SE}_{GM}$ gives the energy which is provided by colliding grinding media without recognizing the stressed product. This stress energy is reduced by the product related energy dissipation term, which was introduced by Becker et al. and can be calculated by the knowledge of the Young´s moduli of product and grinding media [10]. It describes the amount of elastic deformation energy of the grinding media and can be neglected if the product is much softer than the grinding media. The stress energy $\overline{SE}_P$ occurs. The fracturing result of the product is not simply dependent on the stress energy, because often not only one particle is captured and stressed between two grinding media. Thus, the median stress intensity $\overline{SI}_P$ was used, which recognizes the solids content c_V, the particle size x_P and the active volume between two colliding grinding media, which is a function of the grinding media diameter d_{GM} [2].

$$\overline{SI}_P = \frac{\overline{SE}_P}{(d_{GM} \cdot x_P^2) \cdot c_V} \quad \text{Equation 4}$$

This stress intensity reaches maximum values, if optimum grinding conditions are used, thus, the grinding progress is most effective. It represents more or less a value of breakage energy of the product. These optimum stress intensities $\overline{SI}_{P,opt}$ are now correlated with the compression strength of the material σ_{CS} and the number of crystallites within the particles. The grinding boundary is reached when the particles are single crystalline and is called true grinding limit and was introduced by Knieke [11]. The grinding limit x_{min} and the compression strength σ_{CS} can roughly be correlated to each other as figure 2 shows.

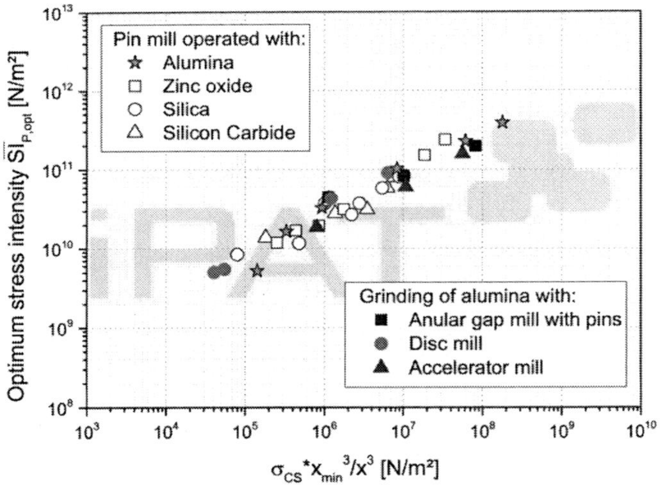

Figure 1: Relation between optimum stress intensity and breakage relevant material parameters

As can be seen from Fig. 1 the correlation keeps its validity for different mills and materials. Therewith it gets possible to predict optimum grinding conditions, if material parameters are known.

$$\overline{SI}_{P,opt} = A \cdot (\sigma_{CS} \cdot \frac{x_{min}^3}{x_P^3})^B \qquad \text{Equation 5}$$

A and B are constant factors. Under the choice of target particle size x_p and the solids concentration c_V, a well suited set consisting of stirrer tip speed, grinding media diameter and density can be chosen in an optimum way.

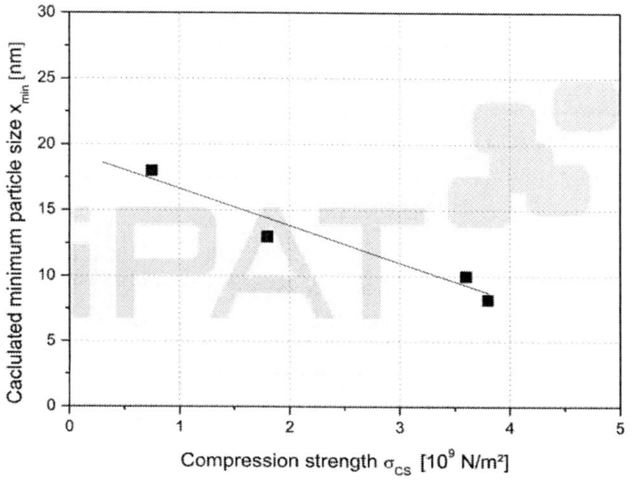

Figure 2: Correlation between calculated minimum particle size x_{min} and compression strength σ_{CS}

Determination of grinding time or energy consumption with the enhanced stress model

The product of stress energy and stress number is proportional to the energy consumption as equation 3 shows. To obtain a certain product fineness a pair existing of stress energy and stress number occurs, as can be seen from Figure 3. The grinding time t_g is a part of the stress number calculation, thus it can be used in the other way around as well. With the knowledge or assumption of the stress number for a certain product fineness and stress energy, the grinding time can be estimated.

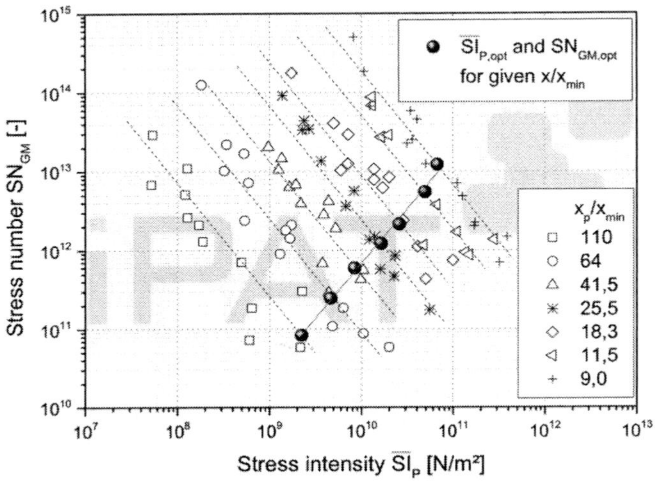

Figure 3: Relation between stress number SN_{GM} and stress intensity $\overline{SI}_P$ for different grinding degrees (Material: Alumina, Mill: Pin mill)

The optimum stress intensities were calculated for the different particles sizes and the stress numbers obtained by using fit functions. As it can be seen a correlation for the optimum cases occurs. This fact was checked for the other mills and materials as well, thus Fig. 4 results.

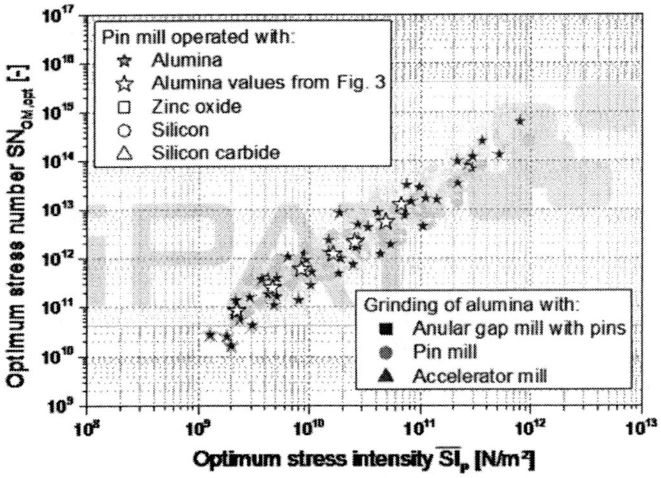

Figure 4: Correlation between the optimum stress number $SN_{GM,opt}$ and the optimum stress intensity $\overline{SI}_{P,opt}$

As can be seen from Figure 4 the optimum stress intensity is also proportional to the optimum stress number.

$$\overline{SI}_{P,opt} = A \cdot \left(\sigma_{CS} \cdot \frac{x_{min}^3}{x_P^3}\right)^B = C \cdot \left(SN_{GM,opt}\right)^D \qquad \text{Equation 6}$$

In the next step a correlation between optimum and non optimum conditions was determined, whereas the slopes at the right and left side of the optimum (original stress model) have to be taken into account. It enables the calculation of the grinding time for chosen operation parameters from the stress number SN_{GM}.

$$\frac{SN_{GM,opt}}{SN_{GM}} = \left(\frac{\overline{SE}_P}{\overline{SE}_{P,opt}}\right)^{\alpha} \qquad \text{Equation 7}$$

The estimation of the energy consumption is only possible if the energy dissipation mechanism in the mill were taken into account. A description of the different energy transfer coefficients is found elsewhere [12]. There are three different energy transfer coefficients. The energy transfer coefficient of the mill υ_{Mill} is a mill dependent values which has to be determined once for each mill, by operating it with water and different operation parameters. The energy transfer coefficient of the suspension υ_{η} is a value which changes with the formulation of the suspension and in some cases also with the operation parameters. Thus, it is a good advice to measure it, when the operations conditions and the suspension formulation are clearly defined. The energy transfer coefficient of the product υ_P can be calculated if the Young´s moduli of product and grinding media are known.

$$E \cdot \upsilon_{Mill} \cdot \upsilon_{\eta} \cdot \upsilon_P = \overline{SE}_P \cdot SN_{GM} \qquad \text{Equation 8}$$

Determination of material parameters with the enhanced stress model

For the here used dolomite no material parameters were provided by the manufacturer. A research for the material parameters gives a more or less wide range, where the parameters might be located as table 1 shows.

Table 1: Material parameters of Dolomite [13]

Young´s Modulus $E_{Dolomit}$	16000 – 80000 N/mm^2
Compression strength σ_{CS}	125 – 250 N/mm^2
Min. particle size x_{min}	17 – 19 nm

With the enhanced stress model and the knowledge of the relations between material parameters, stress intensity and stress number, the estimation of the material parameters from experimental data is possible. Therefor three grinding experiments in the labscale M4IsaMill were carried out, as figure 5 shows. Moreover the stress energy distribution following the approach of Stender was calculated and is shown in figure 6 [14].

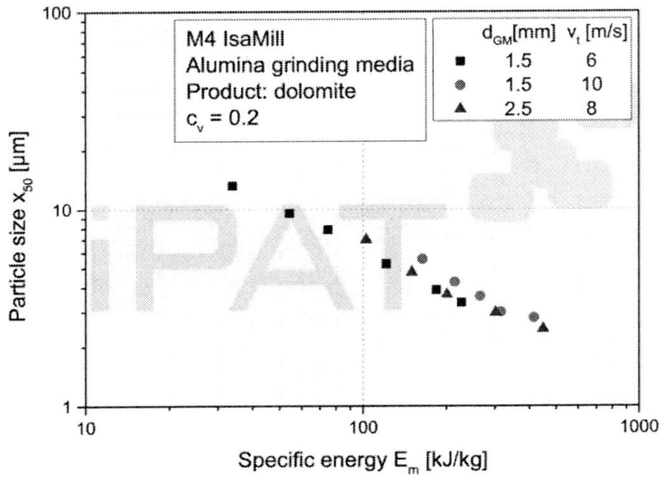

Figure 5: Median particle size as function of the specific energy input for three experiments carried out in the M4 IsaMill

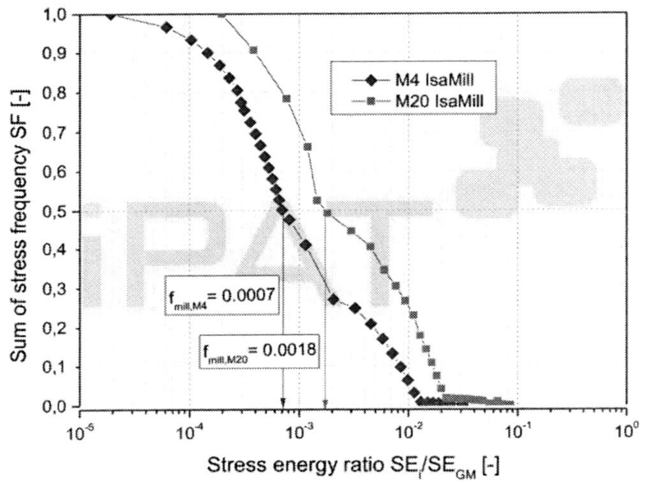

Figure 6: Calculated stress energy distribution for the M4 and M20 IsaMills

Besides the grinding experiments, the mill was operated with pure water to determine the energy transfer coefficient of the mill and the dependency of the stress number or stress frequency to the grinding media diameter. The received power consumption is plotted versus the product of stress frequency and stress energy, whereas the constant $k(d_{GM})$ is adopted in a way, that a good fit occurs. The results for the labscale and the pilotscale mill are shown in figure 6 and have to be carried out once for each mill.

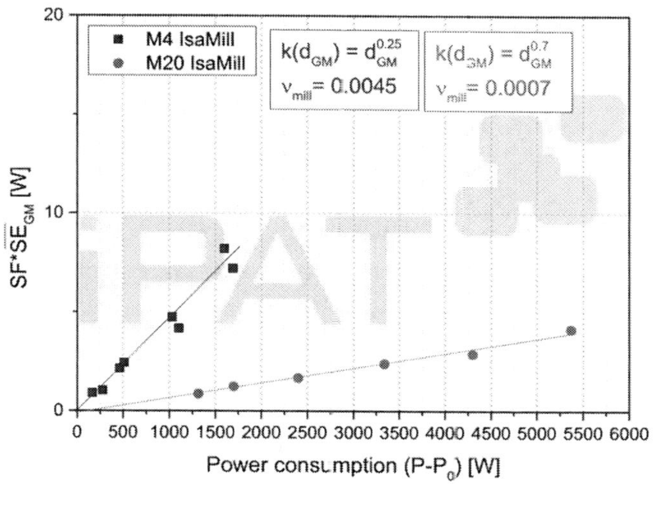

Figure 6: Power characteristic of the M4 and M20 IsaMills

With the knowledge of the constant k(d_{GM}) the stress number SN_{GM} as well as the stress energy $\overline{SE}_P$ for each data point from figure 5 can be calculated. The optimum stress intensity $\overline{SI}_{P,opt}$ and the $SN_{GM,opt}$ optimum stress number were now calculated for different settings regarding table 1 and proved which sets of parameters leads to the best fit function with all the data known from figure 4.

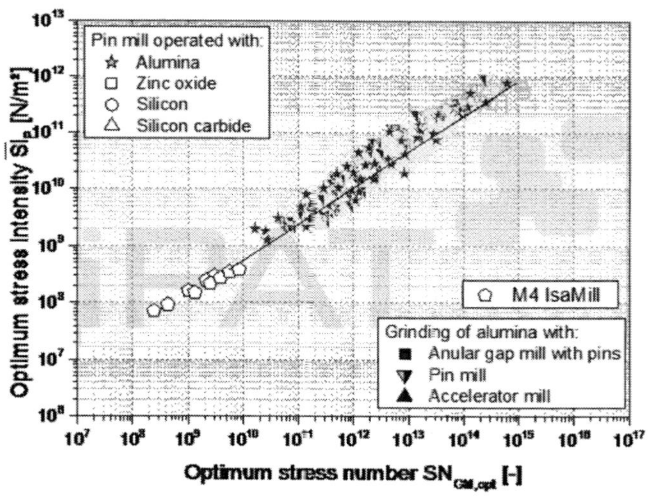

Figure 7: Optimum stress intensity as function of the optimum stress number

Therewith the material parameters were determined and the set-up of the process in the pilotscale mill can be started. The obtained parameters are found in table 2.

Up-scaling the process with the enhanced stress model

For the scale-up of the grinding process of dolomite to the M20 IsaMill, first the stress energy distribution is calculated to obtain the f_{mill}-factor, which is necessary to determine the median stress energy.

At this point it is possible to calculate the grinding times, for a desired product fineness. E.g. if a median particle size of 3 µm is the target, the optimum stress intensity $\overline{SI}_{P,opt}$ can be determined using figure 1 and the material parameters determined in the former part. With the optimum stress intensity on the one hand optimum process parameters can be chosen and the optimum stress number can be figured out by figure 7. Table 2 and 4 summarizes those values.

Table 2: Summary of known or obtained values and parameters

$x_{p=target}$	c_V	ρ_{GM} [kg/m³]	$f_{mill,ISA20}$	$v_{mill,ISA20}$	$k(d_{GM})$	$\nu_{P,Dolo}$	E_P [N/mm²]	σ_{CS} [N/mm²]	x_{min} [nm]	$\overline{SI}_{P,opt}$ [N/m²]	$SN_{GM,opt}$
3 µm	0.2	3600	0.0018	0.0007	$d_{GM}^{0.7}$	0.85	80000	250	17	$3.4*10^8$	$4.8*10^9$

Table 3 Determination of optimum stress conditions

$\overline{SI}_{P,opt}$	$\overline{SE}_{GM,opt} = \frac{\overline{SI}_{P,opt} \cdot d_{GM} \cdot x_P^2 \cdot c_V}{\nu_P}$		$v_t = \left(\frac{\overline{SE}_{GM}}{f_{mill} \cdot d_{GM}^3 \cdot \rho_{GM}} \cdot \right)^{0.5}$	
$3*10^8$ N/m²	d_{GM} = 1.5 mm	d_{GM} = 2.5 mm	d_{GM} = 1.5 mm	d_{GM} = 2.5 mm
	$1.1*10^{-6}$ J	$1.9*10^{-6}$ J	7.4 m/s	4.4 m/s

As operation parameters the next possible combination slightly higher than the optimum should be chosen. In this case the smaller grinding media (1.5 mm) were used at a stirrer tip speed of 8 m/s, because a set up with the bigger ones is not possible. The minimum stirrer tip speed for this mill is 6 m/s. With equation 2 the resulting stress number SN_{GM} can be estimated and therewith the grinding time gets available.

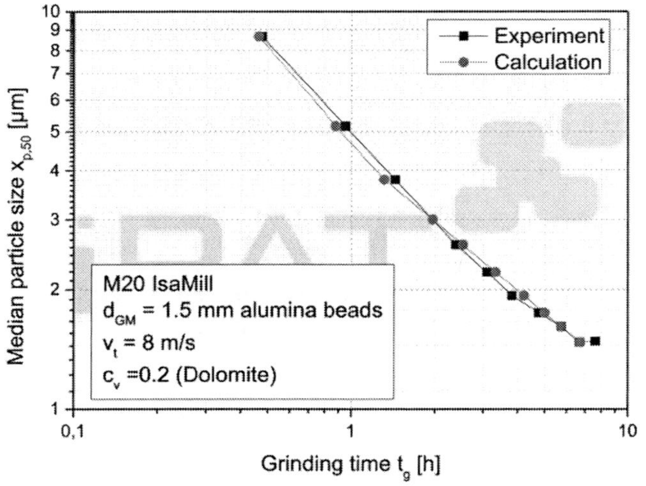

Figure 7: Comparison of predicted and experimental results in relation to grinding time

Now one short experiment is needed to determine the energy transfer coefficient of the suspension v_η. It can be calculated by the relation of the power consumption of the mill operated with water and the power consumption operated with suspension. Here, two experiments were carried out, due to the validation of the model, as figure 9 shows.

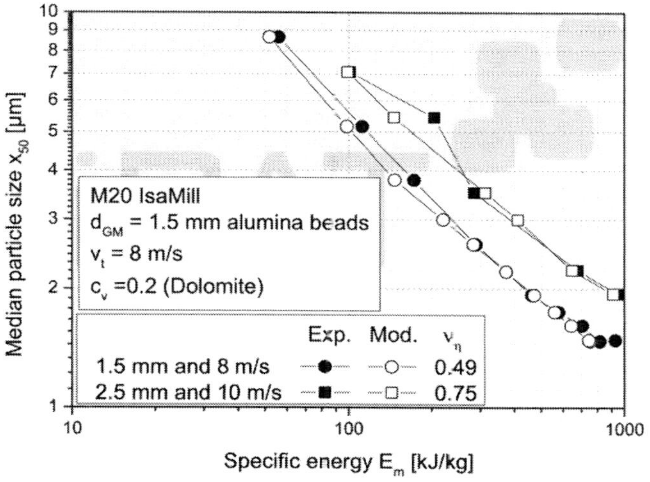

Figure 8: Comparison of predicted and experimental results in relation to specific energy

One further advantage of the usage of the enhanced stress model to set up the grinding process is the possibility to optimize the process in terms of throughput as well. The throughput is calculated by the relation of the power consumption (P-P_0) and the specific energy E_m.

$$Q = \frac{(P-P_0)_{Suspension}}{E_m} \qquad \text{Equation 9}$$

For the two in figure 8 shown experiments, the throughput can directly be calculated, because for the determination of the energy transfer coefficient v_η, the power consumption of the operation with suspension was measured directly during the verifying experiments. In table 4 two other set of parameters are introduced. For one of them the energy transfer coefficient v_η was measured, the other one was estimated under the assumption, that the relation between the energy transfer coefficient v_η and the stress energy to solids concentration relation $(\frac{\overline{SE}_{GM}}{c_v})$ follows a power function (Figure 9).

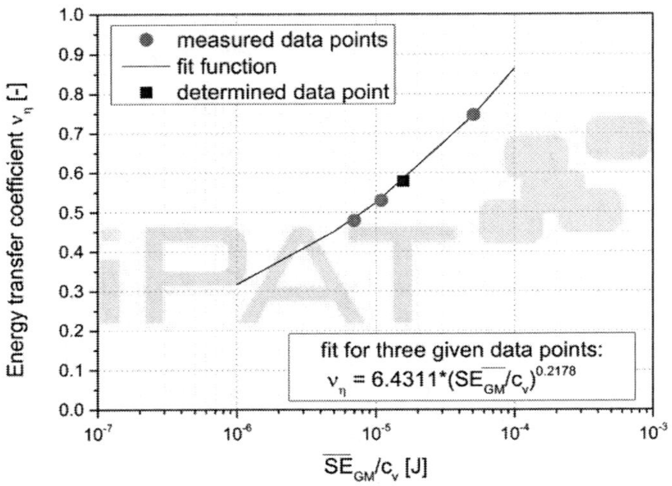

Figure 9: Suspension related energy transfer coefficient as function of stress conditions

Table 4 summarizes the results for the specific energy, grinding time and throughput to obtain a particle size of 3 μm.

Table 4: Summary of grinding relevant data obtained from the model

Set up d_{GM} / v_t	t_g [min]	E_m [kJ/kg]	$(P-P_0)_{Susp.}$ [W]	ν_η [–]	Q [kg/h]
1.5 mm / 8 m/s	118	220	2700	0.49	44.3
1.5 mm / 10 m/s	71	225	4498	0.53	68.8
1.5 mm / 12 m/s	47	259	7192	0.57	100.4
2.5 mm / 10 m/s	122	411	4492	0.75	38.5

The values for the 1.5 mm grinding media with 12 m/s are totally predicted from the model. Results obtained by Schons et al. show a good accordance to the here predicted data [7]. The comparison of the throughputs shows clearly, that the optimization process includes the adaption of the minimum specific energy and the maximum power consumption. Here a compromise of both leads to the highest throughputs.

Conclusion

The grinding of ores and mineral products is often realized in huge stirred media mills with up to several cubic meters of grinding chamber volume. The traditional way of scaling the process is the usage of grinding experiments to obtain optimum process parameters with the original stress model and the determination of Power-number/Re-number-diagrams for the scaling.

Here the enhanced stress model was used to:

1. Determine the material parameters of the used dolomite by three grinding experiments in the lab scale M4 IsaMill.
2. Determine optimum grinding parameters for a desired particle size in the pilot scale M20 IsaMill.

For both points the evaluation of the mill´s "fingerprint" is necessary. This is including the mill factor f_{mill} from the stress energy distribution, the constant $k(d_{GM})$ which recognizes the grinding media dependency within the stress number and the energy transfer coefficient ν_{mill}. The first parameter can be calculated from the mill geometry. For the other two the power consumption of the mill operated with water has to be detected for two or three grinding media sizes/types and different stirrer tip speeds, whereas the tip speed variation is easy to realize during one experiment. Thus, two or three experiments are once necessary to evaluate the mentioned parameters.

The enhanced stress model includes two master curves:

- Relation between product material parameters, particle size and optimum median stress intensity
- Relation between optimum median stress intensity and optimum stress number

These are in the first part used to determine the material parameters of the dolomite, because the master curves work for many different materials and were used to adapt the right set of material parameters for the dolomite, from the grinding kinetics of the material.

With the knowledge of the material parameters the set up the pilot scale M20 IsaMill was realized. Therefor the optimum stress conditions for a desired particle size were found from the master curves. With the known set of parameters the grinding time can be calculated from the stress number. If specific energy and throughput are also desired, the energy transfer coefficient ν_{η} must be measured as well. Therefor the power consumption of the mill operated with suspension and the given set of parameters is necessary, but only short operation times are needed. If this coefficient is determined for minimum three sets of parameters the energy transfer coefficient for other sets of parameters can be assumed by a fit function. Therewith the specific energy and the throughput can be calculated for diverse set of parameters. Which optimization strategy is followed depends on the process owner, because different targets are possible to follow, like e.g. minimum grinding media wear, which is obtained for the minimum specific energy or the maximum throughput.

Acknowledgement

The authors thank gratefully the company Industrie Bitossi for providing the product and the grinding media, as well as for allocating their M20 IsaMill for the experiments in the pilot scale.

Literature

[1] A. Kwade and J. Schwedes, "Breaking characteristics of different materials and their effect on stress intensity and stress number in stirred media mills," *Powder Technology,* vol. 122, pp. 109-121, 2002.

[2] A. Kwade, "A stressing model for the description and optimization of grinding processes," *Chem. Eng. Technol.,* vol. 26, p. 199/205, 2003.

[3] S. Breitung-Faes and A. Kwade, "Design of wet stirred media milling processes via a mechanistic model," in *European Symposium Comminution and Classification,* Göteburg/Schweden, 2015.

[4] S. Breitung-Faes and A. Kwade, "Use of an Enhanced Stress Model for the Optimization of Wet Stirred Media Milling Processes," *Chemical Engineering & Technology,* vol. 37, pp. 819-826, May 2014 2014.

[5] S. Breitung-Faes and A. Kwade, "Prediction of energy effective grinding conditions," *Minerals Engineering,* vol. 43-44, pp. 36-43, 2013.

[6] S. Breitung-Faes, "Estimation of product relating energy of wet operated stirred media mills in terms of process transfer to other mill geometries and sizes," *Minerals Engineering,* vol. 103, pp. 33-42, 2017/04/01/ 2017.

[7] D. Schons, S. Breitung-Faes, and A. Kwade, "Optimizing the grinding of dolomite in the IsaMill," presented at the Comminution '18, Capetown (South Africa), 2018.

[8] A. Kwade, "Determination of the most important grinding mechanism in stirred media mills by calculating stress intensity and stress number," *Powder Technology,* vol. 105, pp. 382-388, 1999.

[9] S. Breitung-Faes and A. Kwade, "Prediction of energy effective grinding conditions," *Minerals Engineering,* vol. 43-44, pp. 36-43, 2013.

[10] M. Becker, A. Kwade, and J. Schwedes, "Stress intensity in stirred media mills and its effect on specific energy requirement," *Int. J. Miner. Process.,* vol. 61, pp. 189-208, 2001.

[11] C. Knieke, "Fracture at the Nanoscale and the Limit of Grinding," Technische Fakultät der Universität Erlangen-Nürnberg, Universität Erlangen Nürnberg, Cuvillier Verlag, 2012.

[12] S. Breitung-Faes, "Estimation of product relating energy of wet operated stirred media mills in terms of process transfer to other mill geometries and sizes," *Minerals Engineering,* vol. 103-104, pp. 33-42, 2017.

[13] M. A. Zürich. Dolomitstein [Online].

[14] H.-H. Stender, A. Kwade, and J. Schwedes, "Stress energy distribution in different stirred media mill geometries," *Int. J. Miner. Process.,* vol. 74S, pp. S103-S117, 2004.

508585 Measurement of Rotational Motion in Granular Couette Flow Using Magnetic Resonance Imaging

Abstract:

Granular rheology is complicated by the dependence of the viscosity upon factors such as solid concentration, shear rate, granular temperature, and inter-particle friction. Many different approaches have been developed within the last three decades, including frictional kinetic theory [1, 2], or non-local granular fluidity models [3]. Testing the validity of these models requires measurements of many different quantities, including the solid volume fraction, mean velocity, fluctuating component of velocity (sometimes called the granular temperature), and the angular velocity variance. Magnetic resonance imaging (MRI) has been used to make local measurements of the solid fraction, shear rate and velocity distribution in opaque systems. Here we extend the technique to characterise the rotational motion of the particles as well.

Granular materials exhibit behaviour resembling all of the three main states of matter, such as yielding in solids, gravity flow in liquids and compressibility akin to gases. These features are the result of the contacts between granules at the scale of the particle diameter. In order to understand industrial and natural granular flows, the dynamics of individual granules must be understood. Discrete element method (DEM) simulations of granular flows are useful since the motions of individual particles are tracked, providing detailed information about the flow characteristics [4, 5]. However, DEM simulations are computationally expensive and are limited to simulating laboratory-scale systems. Therefore, a continuum description for granular materials is necessary to assist in the design of industrial processes and to understand natural phenomena.

In order to develop a continuum model, a rheological model is required to determine the stress distribution within the granular material. In liquids and gases, such a model determines the stress from the local shear rate of the material and its viscosity. In granular materials, the relationship is less clear. Models derived from the kinetic theory of granular flow use an effective viscosity, much like in the liquid or gas phase. These models were developed assuming relatively dilute granular flows. By contrast, models derived from dense systems suggest the stress arises primarily from the confining pressure on the flow. Several experimental and theoretical studies now suggest that rotational motion of the granules should also be considered. In comparison to other variables, particle rotation has not been widely studied, in part due to the scarcity of suitable measurement techniques.

In MRI, pulsed field gradient (PFG) measurements encode for spin position during the acquisition time. From the PFG signal, the local mean velocity and the variance of the velocity fluctuations (i.e. the granular temperature) may be retrieved [6, 7]. Changes in spin position may be caused by translational motion resulting from external forcing of a system and by inelastic collisions between granules. Conventional analysis of the signal assumes that the granules behave as point sources and that there is no contribution to the signal from their rotation. However, spins move along the encoding axis as the particle rotates. Thus, it is expected that the measured velocity fluctuations will be greater than that arising from the translational motion alone.

In this work, we propose a framework by which the angular velocity distribution of a granular material may be measured using MRI. A model is developed to describe the PFG signal in terms of the translational and rotational motion of the granules. The apparent granular temperature measured by MRI, T_{app} is related to the true granular temperature, and the angular velocity variance, by:

(1)

where R is the granule radius. The model was validated against discrete element method simulations of PFG experiments in granular couette flow. The PFG signal acquisition is simulated by modelling each particle as a cluster of points. A simulated MRI signal was then found by numerically integrating the point positions with respect to time. This method offers the advantage of being able to include or exclude the effect of particle rotation upon the signal. Hence, the true granular temperature may be separated from the apparent granular temperature. These simulated measurements were then analysed as in a conventional MRI experiment to determine the mean velocity and fluctuating component of the velocity.

The mean velocity profile was in excellent agreement with physical PFG measurements for the same system. Granular rotation was found to have no effect upon the imaginary component of the signal and hence the mean velocity. Each spin has a counterpart positioned on the other side of the granule with an equal and opposite phase. Thus the net phase for a granule is equal to the phase arising from translational motion only.

Measurements of the fluctuating component of the particle velocity were found to exceed the value obtained from the simulations of the fluctuations of the translational component of motion. Increasing the inter-particle friction coefficient and coefficient of restitution caused the rotational and translational components of the granular temperature to increase, respectively. However, for all contact parameters tested, the simulated granular temperature arising from translational motion was lower than the apparent granular temperature of the PFG experiments. The deviation between the simulation and experimental results could only be removed by including the effects of particle rotation in the calculation of the simulated granular temperature.

Rotational motion increases signal attenuation when measured by MRI, and hence the apparent fluctuating component of the particle velocity is overestimated. A detailed study of the influence of rotation on the signal revealed that the contribution from rotational motion increased with increasing particle radius, as given in the equation above. The simulation results also confirmed the additive relationship between the translational and rotational contributions to the apparent granular temperature. This result provides a means to quantitatively measure the rotational and translational components of the velocity fluctuations.

The decomposition of the measured apparent granular temperature into separate translational and rotational components is not possible using PFG experiments alone. Instead, the true granular temperature must be measured by another experimental method. In this paper we consider using modulated gradient spin echo and dynamic magnetic resonance scattering as possible means of separating both translational and rotational components of the motion. These experiments will provide completely non-invasive measurements of particle dynamics in opaque, particulate flows. Thus, we will show that MRI may be used to fully quantify the granular dynamics and hence probe rheological models of granular flow experimentally.

References:

1. Jenkins, J.T. and C. Zhang, *Kinetic theory for identical, frictional, nearly elastic spheres.* Physics of Fluids, 2002. **14**(3): p. 1228-1235.

2. Lun, C.K.K., *Kinetic theory for granular flow of dense, slightly inelastic, slightly rough spheres.* Journal of Fluid Mechanics, 1991. **233**: p. 539-559.

3. Kamrin, K. and G. Koval, *Nonlocal constitutive relation for steady granular flow.* Physical Review Letters, 2012. **108**(17).

4. Cundall, P.A. and O.D.L. Strack, *A discrete numerical model for granular assemblies.* Geotechnique, 1979. **29**(1): p. 47-65.

5. Kloss, C., et al., *Models, algorithms and validation for opensource DEM and CFD-DEM.* Progress in Computational Fluid Dynamics, 2012. **12**(2-3): p. 140-152.

6. Callaghan, P.T., *Principles of nuclear magnetic resonance microscopy.* 1991.

7. Holland, D.J., et al., *Spatially resolved measurement of anisotropic granular temperature in gas-fluidized beds.* Powder Technology, 2008. **182**(2): p. 171-181.

AUTHOR INDEX

AUTHOR INDEX

AUTHOR INDEX

AIChE
120 Wall Street, FL 23
New York, NY 10005-4020

ISBN 978-1-5108-6970-7
9 781510 869707

CURRAN ASSOCIATES INC.
proceedings.com